INVISIBLE IN THE STORM

INVISIBLE IN THE STORM

THE ROLE OF MATHEMATICS IN UNDERSTANDING WEATHER

IAN ROULSTONE AND JOHN NORBURY

PRINCETON UNIVERSITY PRESS
PRINCETON AND OXFORD

Published by Princeton University Press, 41 William Street,
Princeton, New Jersey 08540
In the United Kingdom: Princeton University Press, 6 Oxford Street,
Woodstock, Oxfordshire OX20 1TW

press.princeton.edu

Library of Congress Cataloging Number 2012037453

ISBN 978-0-691-15272-1

British Library Cataloging-in-Publication Data is available

This book has been composed in Minion Pro

Printed on acid-free paper. ∞

Printed in the United States of America

1 3 5 7 9 10 8 6 4 2

CONTENTS

PREFACE

To most of us, meteorology and mathematics are a world apart: why should calculus tell us anything about the formation of snowflakes? But mathematics has played an ever-growing and crucial role in the development of meteorology and weather forecasting over the past two centuries.

The continuing development of modern computers allows prodigious amounts of arithmetical calculation to be performed every minute, and every day forecasters harness this computational power to predict tomorrow's weather. But to appreciate the success of modern weather forecasting, and to figure out why it occasionally goes wrong, we need to understand how the behavior of the atmosphere and oceans is quantified in terms of mathematics.

Computer operations are couched in mathematical instructions and follow abstract logical rules to organize the calculations, so it is necessary to describe both the present state and the changes of the Earth's atmosphere in terms of appropriate mathematical language for implementation on computers. However, this is an enduring problem for two important reasons. First, we will always have less than perfect knowledge of the interactions of clouds, rain, and the eddying gusts of wind; and second, computers can only execute a finite number of calculations in producing each forecast.

This presents forecasters with an interesting challenge: how to capture the essentials of the behavior of the atmosphere without being "blown off-course" by the lack of perfect knowledge. This challenge preoccupied the pioneers of weather forecasting well before the advent of modern computers. By the end of the nineteenth century, the basic physical laws governing the motion of the atmosphere were in place; attention then turned to finding solutions that predicted the weather.

This book describes the developing role of mathematics in our ongoing quest to comprehend and predict weather and climate. The pursuit of meteorology as an exact, quantitative science was precipitated in the early twentieth century by a relatively small group of mathematicians and physicists from quite disparate backgrounds. Their story is fascinating and informative in itself, and their legacy is more than the foundation of modern weather forecasting—they showed us why the fusion of mathematics and meteorology will always underpin the science of predicting weather and climate.

But along with the realization that forecasting the next storm was a difficult problem in science, mathematicians, while probing the mystery of the stability of the solar system, discovered chaos. The key question was whether the planets would continue their motion around the Sun forever, or whether a chance collision—for example, with a meteor—might eventually lead to their future motion changing entirely.

Today weather forecasters constantly push at extending the limits to predictability that the physical laws encapsulate. Continually improving the forecasting procedure provides more reliability in the predictions and requires bigger computers, ever-better software, and more accurate observations. Within modern supercomputers, the scale of the number crunching almost defies comprehension, but mathematics allows us to see the order amid the detail.

In the first half of the book we describe four key elements in the historic development of weather forecasting: first, how we learned to measure and describe the atmosphere; second, how we encapsulated this knowledge in terms of physical laws; third, how we learned to express the physical laws in terms of mathematics, thereby enabling us to make predictions; and fourth, how we learned to recognize the devil within the detail—the phenomenon we call chaos.

The second half of the book then describes the post-1930s modern approach where the mix of mathematics and machines has enormously improved our ability to predict future weather and climate. The Second World War and the subsequent expansion of civil and military aviation drove many new agendas in meteorology and related technologies such as radar, satellites, and, not least, computers. This technology push facilitated the breakthrough in 1949 of the first weather forecast calculated by a computer. But underlying the well-documented development of this pioneering work lies a little-known story about the role of mathematics in the discovery of the key to successful computer calculations. So we end by describing how math is being, and will continue to be, exploited by forecasters to separate the predictable from the unpredictable. This becomes even more relevant in the context of understanding and predicting future climate.

For scientific completeness, tech boxes are used to detail the technical material in the text. The book is designed so that those who prefer to skip the content of the tech boxes may do so and still understand the overall concepts. The glossary following the postlude uses simple language to explain the concepts used to design weather-forecasting computer programs.

Acknowledgments

Over the many years we have been researching and writing this book, we have become indebted to many friends, family, and colleagues for useful discussions and critical feedback. In this regard, we thank Sid Clough, Mike Cullen, Jonathan Deane, Dill Faulkes, Seth Langford, Peter Lynch, Kate Norbury, Anders Persson, Sebastian Reich, Hilary Small, Jean Velthuis, and Emma Warneford. Particular thanks are due to Andy White for his very careful reading of, and commentary on, the penultimate draft of the manuscript.

We also thank Sue Ballard, Ross Bannister, Stephen Burt, Michael Devereux, David Dritschel, Ernst Hairer, Rob Hine, Rupert Holmes, Steve Jebson, Neil Lonie, Dominique Marbouty, John Methven, Alan O'Neill, Norman Phillips, David Richardson, Claire Roulstone, and Mike White for their help in providing artwork and illustrations.

We are grateful to the following organizations that provided images, and in particular we would like to thank staff at the American Meteorological Society, the University of Dundee, the European Centre for Medium-Range Weather Forecasts (ECMWF), the Met Office and National Meteorological Library, and the Royal Society of London, for valuable help. Ian Roulstone gratefully acknowledges the support of the Leverhulme Trust, via a fellowship in 2008–9.

We would like to thank Vickie Kearn of Princeton University Press for her patience and constant encouragement, and her many colleagues, past and present, including Kathleen Cioffi, Quinn Fusting, Dimitri Karetnikov, Lorraine Doneker, Anna Pierrehumbert, Stefani Wexler, and Patti Bower, for their help in producing the book.

Finally, the love and support of our families was, to say the least, invaluable, and we dedicate this book to them.

PRELUDE

New Beginnings

By the end of the nineteenth century mankind was using Newton's laws of motion and gravitation to calculate the times of sunrise and sunset, the phases of the Moon, and the tides. Such data was carefully tabulated in almanacs and diaries, and many working people, from fishermen to farmers, benefitted from this successful application of science. Then, in 1904, a Norwegian scientist published a paper outlining how the problem of weather forecasting could be formulated as a problem in mathematics and physics. His vision became the cornerstone of modern weather prediction, and the agenda he pursued for the next three decades stimulated many talented young scientists—their research laid the foundations of contemporary meteorology.

From the outset, no one was under any illusion about how difficult it would be to calculate the weather. Predicting the winds and the rain, the fine and the dry spells—by working out the variations in air pressure—temperature, and humidity around the entire planet was recognized as a problem of almost immeasurable complexity.

This is the story behind the intellectual journey to understand and predict the ever-changing weather using physics, computers, and mathematics. Mathematics is important not only as the language that defines the problem but also because it provides solutions on modern supercomputing facilities. Further, we use math to get more information from the computer output, which we then use to make the forecast. The forecasts influence decisions in everyday life—from whether to take an umbrella to work, to how to design flood defenses to protect communities for decades to come.

Images of the Earth from space changed many people's view of our home. We notice the universality of water in its various states, from the oceans and ice sheets, the clouds and rain, to that locked up in all life. Cloud patterns show the winds that transport heat away from the tropics, keeping the ice sheets at bay. Could the weather systems change, and if they did, what would be the consequences for the distribution of water and life in its various forms? In this book we focus on the Earth's atmosphere and explain how mathematics enables us to describe and predict the endless cycle of weather and climate.

Figure Pr.1. This Blue Planet image is taken from the NASA website, and shows clouds swirling around the Earth. Is it possible, using mathematics, to calculate how these cloud patterns change over the next five days, and how the Arctic ice-sheet changes over the next five years? Reproduced courtesy of NASA.

ONE

The Fabric of a Vision

Our story begins at the end of the nineteenth century in the twilight years of the theory of the "ether": a theory of space, time, and matter, which was soon to be superseded by Einstein's theory of relativity and by the theory of quantum mechanics. A Norwegian scientist made a remarkable discovery while working on the ether theory, a discovery that was to lead to a new beginning in meteorology.

A Phoenix Arises

A pensive, thirty-six-year-old Vilhelm Bjerknes peered through a quarter-pane of his window at a city shrouded by a sky as gray as lead. It was a bitterly cold afternoon in November 1898 and Stockholm was preparing itself for a taste of winter. Snow had been falling gently since early morning, but a strengthening northerly wind began to whisk the flurries into great billows that obliterated the skyline. We may imagine Bjerknes returning to the fireside with anticipation of the warmth keeping the chill at bay, relaxing in a chair, and allowing his thoughts to wander. As the fire began to roar and the blizzard strengthened, his growing sense of physical comfort was accompanied by a feeling of inner contentment: he was at one with the world. However, this wasn't just the simple pleasure that comes from finding sanctuary from the winter's rage; it was deeper and much more profound.

Bjerknes gazed at a spark as it flickered and swirled up the chimney. The tiny cinder disappeared from view, carried into the even greater swirl of the wind and snow outside. He continued to watch the dance of

Figure 1.1. Vilhelm Bjerknes (pronounced Bee-yerk-ness), 1862–1951, formulated weather prediction as a problem in physics and mathematics.

the smoke and flames, and listen to the howl of the storm. But he did so in a way that he had never done before: the spiraling of smoke above the fire, and the intensification of the storm—events that had been witnessed by mankind since the dawn of civilization—were two manifestations of a new theorem in physics. It would amount to a small landmark—not quite as prominent in the timeline of science as Newton's laws of motion and gravitation—but all the same it would explain fundamental features of weather. The salient idea had remained hidden, locked away from meteorologists behind the heavy door of mathematics. The new theorem was due to Vilhelm Bjerknes, but it was destined to do much more than carry his name; it would propel meteorology into a cutting-edge science of the twentieth century and pave the way to modern weather forecasting. And it would do this because, above all else, it would first change his vocation.

But the irony was that Bjerknes never intended his ideas to shape history, or his own destiny, in this way at all. Indeed, while he relished the excitement of opening a new window onto the laws of nature, he agonized over his priorities and ambitions, and he began to question the future of his career. His newfound vision was borne out of a rapidly waning and increasingly unfashionable development in theoretical physics. For more than half a century, a group of leading physicists and mathematicians had been trying to decide if phenomena such as light and forces such as magnetism travel through empty space or whether they travel through some sort of invisible medium.

By the 1870s there was a growing consensus that empty space must in fact be filled with an invisible fluid, which was called the ether. The basic idea was very simple: just as sound waves travel through the air, and just as two boats passing each other feel their mutual presence because the water between them is disturbed, light waves and magnetic forces should travel through some sort of cosmic medium. The scientists trying to understand and quantify the properties of such an ether believed that there must be some similarity to the way water, air, and other fluids affect and are affected by objects that move within them. By showing how experiments with objects immersed in water replicate the type of effects that are familiar from experiments with magnets and electrical devices, they conceived of demonstrating the existence of this ether.

In 1881, at the prestigious Paris International Electric Exhibition, which attracted the likes of Alexander Graham Bell and Thomas Alva Edison, a Norwegian scholar by the name of Carl Anton Bjerknes, a professor of mathematics from the Royal Frederick University in Christiana (now called Oslo), and his eighteen-year-old son, Vilhelm, exhibited their experiments aimed at verifying the existence of the ether. Observers, who included some of the most outstanding scientists of the times such as Hermann von Helmholtz and Sir William Thomson (who became Lord Kelvin), were clearly impressed. Bjerknes and his son won a top accolade for their exhibit, and this success placed them firmly in the spotlight of the international physics community.

Their rising fame and status inevitably led to the gifted young Vilhelm following in his father's footsteps, not only as a mathematician and physicist but also as one of the proponents of the theory of the ether. Research on this hypothesis was refueled in the late 1880s when Heinrich Hertz, in a series of extraordinary experiments, demonstrated the existence of electromagnetic waves propagating through space, as predicted by the Scottish theoretical physicist James Clerk Maxwell. In 1894 Hertz published (posthumously) a book outlining his ideas for how the ether should play a crucial role in formulating the science of mechanics.

Now this was no small undertaking. We are taught that the science of mechanics was born when Galileo introduced the concept of inertia, and Newton quantified the laws of motion by relating force to acceleration, and so on. We court triteness to mention the success of mechanics in describing the motion of everything from ping-pong balls to planets. But Hertz believed there was something missing; that is, the great bastion of Newtonian mechanics appeared to rely on some rather intangible concepts. So he set out, in an axiomatic way, a general strategy for explaining how actions within the ether could explain phenomena that hitherto required the more elusive ideas of "force" and "energy" that appeared to influence our world without any tangible mechanism for doing so. The general principles set forth in Hertz's book appeared to systematize the program Vilhelm's father had initiated. Carl Bjerknes's work lacked any underpinning rationale, but the Hertzian thesis promised to change all that and, in so doing, would vindicate his life's work. This was an important motivation for his son; Vilhelm, captivated by Hertz's profound ideas, decided to devote his energies to this worldview.

Vilhelm also realized that success with this program would place him at the forefront of physics—an attractive prospect for a determined and ambitious young scientist. The nineteenth century had already seen some remarkable marriages of ideas and theories. In a paper published in 1864 Maxwell unified electricity and magnetism, two hitherto apparently unrelated phenomena. The concepts of heat, energy, and light had also been placed on a common basis, and Vilhelm envisaged that this process of unifying seemingly disparate parts of physics would continue until the entire subject lay on the sure foundations of mechanics—a "mechanics of the ether." He alluded to this vision in his defense of his thesis in 1892, at the age of thirty. Two years later, with his ideas vindicated by Hertz, he embarked on the road to fulfilling this dream.

Bjerknes's work had already taken him to the point of rubbing shoulders with others who sought a unified view of nature via the existence of the ether. One such person was William Thomson, Baron Kelvin of Largs. Thomson was born in Belfast in 1824 and moved to Glasgow in 1832. As a teenager, his curriculum vitae made impressive reading. He took courses at Glasgow University at the age of fourteen, continuing his education at Cambridge University at the age of seventeen. On graduating from Cambridge he spent a year in Paris engaging in research with some of the outstanding mathematicians and physicists of the era. Thomson then resumed his career in Glasgow where, at the age of twenty-two, he was appointed as a full professor to the Chair of Natural Philosophy. Although he was primarily a theoretician of the highest caliber, he had significant practical abilities, which were to create the basis of his considerable wealth. He divided his time between theoretical physics and making money from his expertise in telegraphy: he patented what was later to become the standard receiver used in all British telegraphy offices. Thomson was knighted in 1866 for his work on the transatlantic cable, which transformed communication between Europe and the United States, and fairly soon between other countries around the world. This achievement also facilitated rapid communication between meteorological observers. In the United States he was made a vice president of the Kodak company, and back home his achievements were honored by the public for a second time with an elevation to the peerage in 1892, whereupon he gained the title of Lord Kelvin.

So Kelvin (as he is usually known) was wealthy and held high office—indeed, he was one of the first to make a huge success of combining an academic career with industry—but he was undoubtedly preoccupied with the one thing that had created his good fortune: science. His contribution to theoretical physics was enormous. Kelvin played a leading role in explaining heat as a form of energy when he came to support the

Figure 1.2. Sir William Thomson, later Lord Kelvin (1824–1907), a distinguished professor at the age of twenty-two, continued to dominate science for the rest of his life. He published more than six hundred papers and was serving a third term as president of the Royal Society of Edinburgh at the time of his death. Kelvin made a considerable fortune from his work on the transatlantic cable: he bought a 126-ton yacht, the *Lalla Rookh*, as well as a fine house in the Scottish coastal town of Largs. He is buried in Westminster Abbey, next to Sir Isaac Newton.

somewhat radical and abstract view for its day, that energy, not force, lay at the heart of Newtonian mechanics. Indeed, the concept of energy would ultimately lie at the heart of science. He was also an enthusiastic supporter of the concept of the ether; in fact, his views on the role of this entity went much further than those of many of his colleagues.

While studying the basic equations of fluid mechanics—Newton's laws applied to the motion of liquids and gases—Kelvin had become particularly interested in a result that had been published in 1858 by Hermann von Helmholtz. In his analysis of fluid motion, Helmholtz conceived of the idea of a "perfect fluid"—a liquid or a gas that is assumed to have some very special properties. The term "perfect" alludes to the idea that the fluid is assumed to flow without any resistance or friction, hence, no loss of energy into heat or anything similar. Although such a concept is artificial—a product of good old ivory-tower academe—Helmholtz's analysis of its motion revealed something quite remarkable. Instead of analyzing motion in terms of the speed and direction of the fluid flow (the fluid velocity), he studied an equation for the change in the vorticity, or "swirl," of the fluid caused by a "perfect eddy"—that is, one that rotates like coffee in a cup stirred uniformly. To his amazement, this equation showed that if the fluid possessed swirl to begin with, then it would continue to swirl forever. Conversely, if the perfect fluid did not possess swirl, then it would never spontaneously begin to swirl.

Kelvin sought to transfer these ideas and interpret them in the context of the ether. He envisaged the ether as a perfect fluid, and matter as composed of "vortex atoms." That is, he imagined that tiny vortices in the ether were the very building blocks of matter. In 1867 he published (under the name William Thomson) an eleven-page paper entitled "On Vortex Atoms" in the *Proceedings of the Royal Society of Edinburgh*. Naturally, the question arose as to how the vortices, or his "atoms of idealised eddies," had been created in the first place, given that they must be permanent features of a perfect fluid. To appreciate Kelvin's answer to this question, we should remember that the mid-1800s was a turbulent time for science and society; a considerable wake had been created in philosophy and religion by Darwin's ideas on evolution, and this had resulted in rifts between science and the Church. Kelvin, as a devout Presbyterian, saw a way of healing some of these wounds if the creation of the vortices in the ether could be attributed to the hand of God. So, with some exact mathematics on one hand and a clear role for God on the other, Kelvin became committed to the view that the ether was at the heart of all matter and therefore of all physics.

But true to the form that had led him to patent his "phone calls across the Atlantic" ideas, Kelvin did not confine his interest in the ether to wide-eyed speculation. He took Helmholtz's theory of vortex motion and reexpressed it as a theorem that showed how a quantity that measures the strength of vortex motion, which is called circulation, is *unchanging* as the flow of fluid changes. Since circulation plays a major role in understanding weather, we return to discuss it more carefully in chapter 3. For now, we just say that the circulation of an ideal circular eddy (or whirlpool) is the magnitude of the swirling velocity multiplied by the circumference of the circle that the fluid flows around.

Although the ether theories were ultimately doomed, Kelvin's intuition to focus on the circulation in a fluid was of major importance—Kelvin's theorem is a main result in present-day university courses on ideal fluid flow. Bjerknes tried to use these ideas to explain some of his recent experimental results. He had been studying what happens when two spheres are immersed in a fluid and set spinning. Depending on their relative motion, the spheres will either attract or repel each other due to the motion they create in the fluid. Bjerknes was analyzing this theory and trying to show how it might explain forces such as magnetism; however, it was not long before he ran into trouble. His experiments and calculations indicated, contrary to the results of Helmholtz and Kelvin, that vortices might be created in a perfect fluid when spheres were set in rotation adjacent to one another.

This conundrum vexed Bjerknes for some time: Kelvin's result was mathematically sound, so how could the results he obtained from his own experiments be wrong? In the meantime, Bjerknes had obtained a position at the new Swedish högskola in Stockholm. The privately funded university held pure research in the highest regard, and Vilhelm's new position offered him plenty of opportunity to pursue his interests. One day in early 1897, while walking home from the högskola, he realized that Kelvin's theory, while correct, would not apply to the experiment (and problem) he was studying. The key fact was that Kelvin's circulation theorem made no allowance for the possibility that pressure and density might vary independently of one another in the fluid, as they do in the atmosphere. Such variations in Bjerknes's experiment would invalidate the application of Kelvin's theorem. Bjerknes immediately set about trying to modify Kelvin's theory to allow for independent variations in the pressure and density, and he succeeded in showing how circulation might be created, strengthened, or weakened, even for perfect fluids. The result is known as Bjerknes's circulation theorem.

This was a major breakthrough. By the end of the nineteenth century, mathematicians and physicists had known for nearly 150 years how to apply Newton's laws of motion to study and quantify the dynamics of fluids. The problem, however, was that the equations involved, even for perfect fluids, were horrendously difficult to solve, and by the late 1880s only a few very special and idealized solutions had been found. By concentrating on vorticity and circulation rather than on the speed and direction of fluid flow, Helmholtz and Kelvin opened the door to "back of the envelope" (or very simple) calculations to work out how vortices—ubiquitous features of fluid motion—move and change with time. Calculations that would be hugely complicated if we had to work out how the speed and direction of all the particles of the flow would change directly from the laws of motion were now reduced to a few relatively easy steps.

Bjerknes extended these ideas, and his theorem enables us to deal with much more realistic situations than those covered by the work of Helmholtz and Kelvin. In particular, this new theory allows us to work out how vortices in the atmosphere and oceans behave because in these nearly perfect fluids pressure, temperature, and density are all interconnected. Consequently, without even attempting the impossibly complicated task of solving the basic laws of motion for the details of the entire fluid, we get a holistic view of how these vortex patterns will morph and change with time. In turn, this helps us to explain the basic swirling structure of weather systems, which are often most obvious in pictures taken from satellites (such as that in figure 1.3, and in figure CI.2 in the color section, which shows similar eddies in the Gulf Stream). The persistence of these large swirling eddies of air and clouds in the atmosphere is an example of Kelvin's theorem at work, modified by the actual temperature and density variations as proposed by Bjerknes. In the Gulf Stream, the saltiness (or salinity) of the water acts to change the density, and so modifies its eddy behavior in a similar fashion.

Bjerknes was overjoyed with his advances and began to discuss his findings with his colleagues. In late 1897, before the Stockholm Physics Society, he presented his generalizations of Helmholtz and Kelvin's theories. It was at this point that serious interest in his work was aroused,

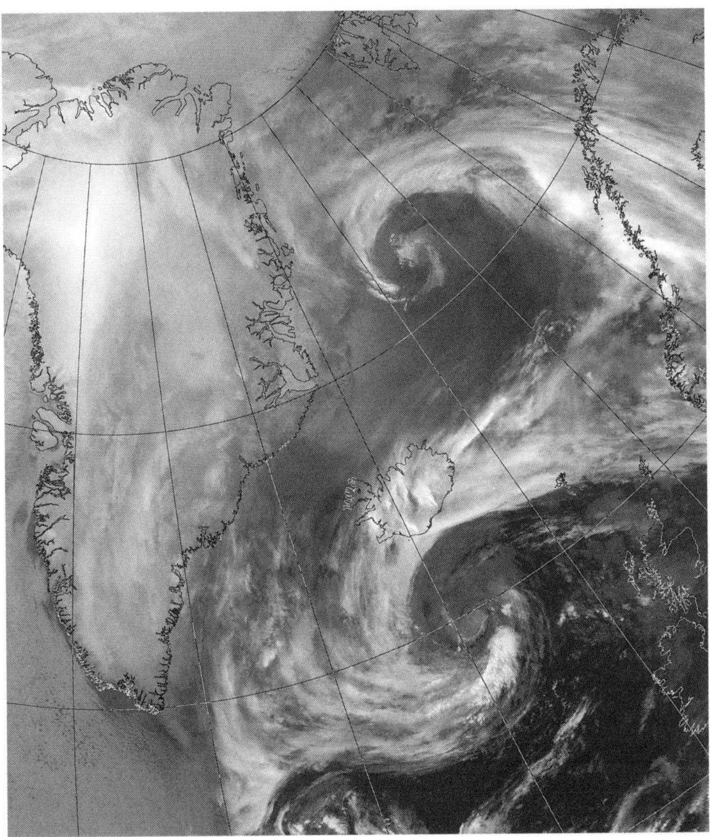

Figure 1.3. A cyclone, or low-pressure system, is a large rotating mass of air, about one thousand kilometers in diameter, that moves through our atmosphere changing the weather. These weather systems often bring rainfall and squally winds. The name cyclone comes from the Greek for the coiling of a serpent. © NEODAAS / University of Dundee.

not from among the dwindling supporters of the theories of the ether but from scientists in very different, important, and emerging areas of physics. Svante Arrhenius, a fellow member of the Physics Society, was interested in applying such ideas from mainstream physics to problems that were becoming particularly well defined in atmospheric and oceanic sciences. He was a distinguished chemist and was among the first to discuss the greenhouse effect of carbon dioxide.

Both the predictable and the unpredictable features of the oceans and the atmosphere have stimulated scientific thinking since science began, but serious scrutiny of atmospheric science proved difficult, and only a few individuals working mainly in isolation had managed to contribute to the subject in any substantial way before the end of the nineteenth century. One of the reasons for the lack of solid progress in atmospheric science compared to astronomy in the eighteenth and nineteenth centuries was the intractability of the mathematics involved in solving the equations that describe fluid motion. By the late nineteenth century, few scientists had even begun to think about the motion of the atmosphere in terms of a problem for mathematical physics. It was just too difficult!

In Stockholm, Nils Ekholm, a meteorologist who was interested in the formation of cyclones (one of the major discussion points in meteorology since 1820), became a close colleague of Bjerknes. Ekholm had studied the development of these cyclones, the swirls of air visible in figure 1.3, by plotting charts showing how air pressure and air density vary from one location to another. However, he lacked any method for linking his observations to a theory of dynamics until Bjerknes started to talk about his work. Another of Ekholm's interests was ballooning, for which it is necessary to estimate the state of the weather at a given altitude because balloons have to navigate aloft. At that time, very little was known about the structure of our atmosphere at different altitudes (such as how temperature varies with height), but there was an eagerness to acquire that knowledge.

Balloons featured prominently in many intrepid adventures. The Norwegians and Swedes had an enviable record of polar exploration. Fridtjof Nansen, the noted Scandinavian explorer, captured the world's imagination when he traveled by ship around the Polar Sea to the north of Russia and Siberia in the early 1890s. So in 1894 the idea was put forward to reach the North Pole by balloon. Three years of planning and fundraising—which won the financial support of the King of Sweden and Alfred Nobel (of the Prizes fame)—led to an attempt in July 1897. Ekholm was involved in the preparation and in forecasting the weather for the trip, but he did not join the crew. This was very fortunate for Ekholm because the balloon, with its crew of three, disappeared (see figure 1.5). This tragic loss was a great blow to Scandinavian national pride. A rescue mission was considered, but no one knew where to search. Because there was scarcely any knowledge of the winds aloft, any clues as to where the balloon might have gone were down to pure guesswork.

Figure 1.4. The British meteorologist James Glaisher made several balloon ascents in order to study the upper atmosphere. During one ascent in 1862 he and his pilot, Henry Coxwell, reached ten thousand meters—and very nearly died in the process. The freezing temperatures, combined with tangled rigging between the basket and the balloon, made it difficult to operate the balloon's control valve. The balloon was ascending rapidly and Glaisher lost consciousness in the thin air. Coxwell began to suffer from frostbite and he could not use his hands to pull on the cord to release the valve. Eventually he managed to scramble out of the basket, into the rigging, and caught the cord between his teeth and opened the valve. The balloon began to descend and Glaisher regained consciousness. On landing safely, Glaisher walked more than eleven kilometers to find help to recover the balloon and their equipment.

On hearing Bjerknes's lecture in late 1897, Ekholm must have felt like shouting "eureka!" It dawned on him that information about winds and temperature in the upper atmosphere could be obtained by using the circulation theorem together with readings of pressure and wind velocity at ground level, and such information might very well have enabled the balloonists to be rescued. Ekholm had understood that Bjerknes's theorem explained the circulation of entire swirling masses of air, such as cyclones and anticyclones. Because the circulation theorem is based on a formula that relates the strengthening or weakening of circulation to the varying patterns of pressure and density, the structure of cyclones could be estimated. Consequently, observations of a small part of that

Figure 1.5. Solomon Andrée and Knut Fraenkel with the crashed balloon on the pack ice; photographed in 1897 by the third member of the team, Nils Strindberg. This picture was developed from the film recovered in 1930, long after the crew had perished.

mass—such as the air near the ground—would tell us how the air must be moving in the regions we cannot observe directly (assuming the air motion is such that the circulation theorem can be applied). Ekholm lost no time in discussing possible applications of the circulation theorem with Bjerknes, and these discussions marked the beginning of Bjerknes's change in vocation, away from the doomed theories of the ether and toward meteorological science. This scientific rebirth not only changed Bjerknes's life, it was also to shape modern meteorology. Air travel was to develop rapidly from balloons and dirigibles to fixed-wing biplanes in the early twentieth century; methods based on Bjerknes's ideas would enable the air currents of the lower, middle, and upper atmosphere to be understood.

In February 1898 Bjerknes presented another lecture to the Stockholm Physics Society in which he outlined ideas for using the circulation theorem to test Ekholm's conjectures about the formation of air currents in cyclones. Bjerknes remarked that, with the aid of information from balloons and kites that carry instruments to measure winds and temperatures aloft, his theory might be able to resolve long-standing debates about the anatomy of cyclones. He went on to demonstrate how the theory could have been used to guide a search party toward the stricken balloon seven months earlier. This lecture fired the audience with enthusiasm. The Society offered funds to pay for the balloons and kites needed to make the observations Bjerknes suggested. In fact, they established a committee to look into the construction of "kites and flying machines powered by electric motors" to gather the data required, and Bjerknes's long-standing comrade Arrhenius wanted to include an account of the circulation theory in a book he was writing on the latest advances in "cosmical physics."

For Bjerknes it might have seemed that things couldn't get much better. But they did. Another early supporter and beneficiary of his ideas was a professor of chemistry in Stockholm, Otto Petterssen. Petterssen was interested in oceanography and in particular in the patterns of salinity and temperature observed in the oceans. He was aware of the new opportunities for "technology transfer" between mainstream physics and meteorological sciences through Bjerknes's ideas, and the next role for the new theorem was motivated by the depletion of fish stocks. Petterssen had been studying surveys of the waters off the Swedish west coast, which had been commissioned in the 1870s following the unexpected return of herring shoals after an absence of nearly seventy years. When the fish disappeared from the area, the local community fell into economic ruin, and no one knew why or where the fish had gone. The surveys revealed that water with a relatively high salinity and temperature was flowing through the region, and the herring followed such currents. But when cooler, less saline water flowed into the region, the fish would disappear. The question thus arose as to whether it might be possible to predict when this would occur, and it turned out that the Bjerknes's circulation theorem could help provide the answer.

Encouraged by this enthusiasm to apply his new ideas, Vilhelm presented a third lecture to the Stockholm Physics Society in October 1898. In that lecture he established the role of the circulation theory in a wide range of problems, from the flow of heated air up a chimney, to the calculations of the strengthening and weakening of cyclones, and to the prediction of currents in the ocean. But amid this newfound success, Vilhelm was forced to concede that his original ambition of placing his father's work at the heart of modern theoretical physics might be untenable. While the pioneers of modern meteorology and oceanography listened to him intently, the mainstream theoretical physicists were slowly

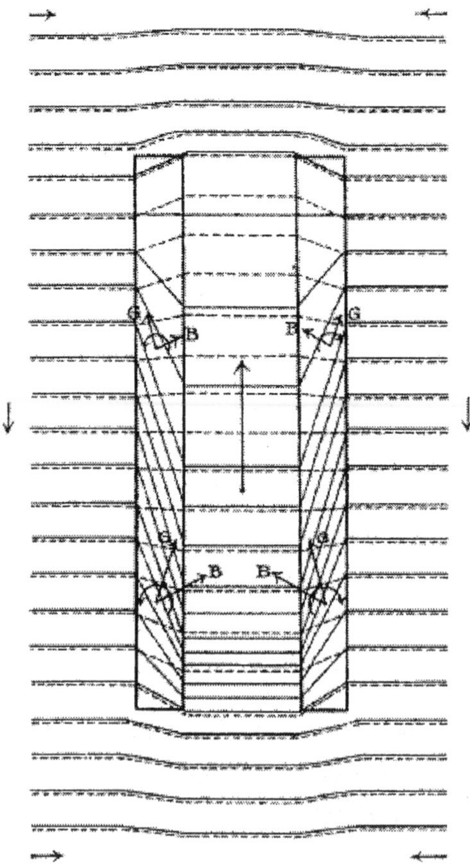

Figure 1.6. Reproduction of figure 7 from Bjerknes's 1898 paper showing the flow of warm air up a chimney. See tech box 1.1 for an explanation of the ideas relating heat, density, and pressure that are encapsulated in this sketch. Bjerknes's imagination stretched from heated air rising up a chimney to raging winter storms. *Bulletin of the American Meteorological Society,* 84 (2003): 471–80. © American Meteorological Society. Reprinted with permission.

but surely turning their backs on any theory of the ether. The turn of the century would usher in a new era: theories of relativity and quantum physics would revolutionize our view of space, time, and matter, and would dispense with the need for an ether altogether.

So Vilhelm Bjerknes stood at one of life's crossroads. The reality, which he had yet to fully appreciate, was that the intellectual triumph of his circulation theorem was like a phoenix rising from the ashes of a spent idea, an idea to which his father had devoted his entire career only to be faced with the grim reality that his life's work was destined for oblivion. As Bjerknes continued to watch the flames in his fire and listen to the storm on that cold November evening in 1898, he was gradually becoming aware of the possibilities that lay before him as the proponent of a new application of physics: not in explaining the invisible,

Tech Box 1.1. Circulation and the Sea Breeze

Bjerknes's vision was that, using measurements of pressure and density (or temperature), we could predict average winds even when locally complicated conditions caused lots of varying gusts and eddies. In this tech box we outline the mathematics of his circulation theorem as it applies to the creation of a sea breeze near a coast warmed by the sun.

The circulation is defined by a path, or contour integral $C = \int \mathbf{v}.d\mathbf{l}$. Here $\mathbf{v}$ is the wind vector and $d\mathbf{l}$ is a small piece of the path around which the integral is performed. Typically, such a path begins over the sea and extends for thirty or so kilometers over the shore and onto the land, from which it returns in the upper atmosphere, as shown by the path arrows in figure 1.7. This integral adds up the total component of the wind that blows in the direction of the chosen path or contour, where on each piece of path $d\mathbf{l}$, $\mathbf{v}.d\mathbf{l}$ is the path component of the wind multiplied by the length of that piece of path.

According to Kelvin's simpler version of the circulation theorem, the rate of change of circulation as measured by an observer moving with the flow is zero. In symbols, $DC/Dt = 0$, where the material derivative D/Dt denotes the rate of change of some quantity that is defined on a fluid parcel as that parcel moves. Bjerknes's contribution was to note that $DC/Dt \neq 0$, that is, circulation can be generated or destroyed when there is nonalignment of surfaces of constant pressure and density in the fluid caused by varying density or temperature.

Pressure surfaces in the atmosphere are nearly horizontal—layers of air of differing temperature and density press down on the layers below because of their weight caused by the Earth's gravity. So the major component of the pressure is due to this weight effect. Bjerknes, in his 1898 paper, looked for applications where heating in the atmosphere changes the density of the air more significantly than the pressure. Thus we show a figure from Bjerknes's 1898 paper below, where the Sun has warmed the coastal plain more than the adjoining ocean, causing the surfaces of constant density—shown

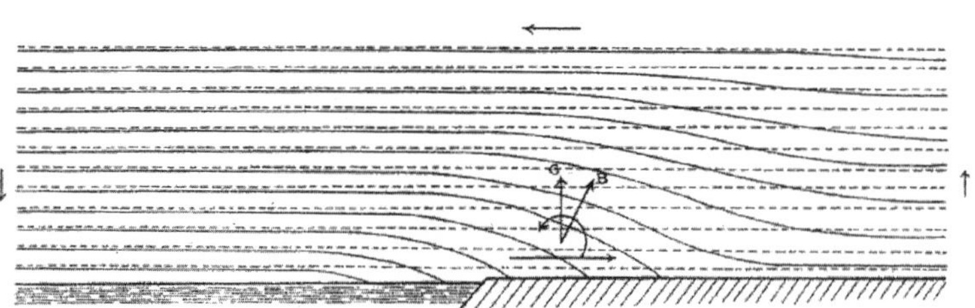

Figure 1.7. Figure 9 from Bjerknes's 1898 paper. Here we see an idealized (daytime) sea breeze forming at the coastline, shown with the sea to the left and the (warmer) land to the right. *Bulletin of the American Meteorological Society* 84 (2003): 471–80. © American Meteorological Society. Reprinted with permission.

as solid lines—to dip (relatively) over the land because the air above the coastal plain has been heated more and is less dense since it has expanded more. The dashed lines show surfaces of constant pressure, and we note that these surfaces are no longer parallel to the surfaces of constant density over the coastal region.

Bjerknes made good use of vectors, denoted here by the arrows **G** and **B** (they are also used in figure 1.6 from Bjerknes's 1898 paper). The vector **B** is perpendicular to the density surface while the vector **G** is perpendicular to the pressure surface. Bjerknes had shown that $DC/Dt = \iint(\mathbf{G} \times \mathbf{B}) \cdot \mathbf{n}dA$, where **n** is the unit vector perpendicular to the plane containing the area element dA and A is the area enclosed by the contour. This new integral is the net sum of misalignment $\mathbf{G} \times \mathbf{B}$ inside the contour path. When **G** and **B** are aligned, $\mathbf{G} \times \mathbf{B} = 0$, which means $DC/Dt = 0$, and Kelvin's theorem applies. So Bjerknes set about formulating a theorem to deal with the misalignment. When **B** and **G** are no longer aligned, circulation is created, hence a sea breeze blows onto the land.

In spite of the many small gusts and eddies in coastal breezes, the creation of overall circulation means that an overall onshore wind is set up—the ubiquitous sea breeze. Bjerknes's idea was that systematic nonalignment of his **B** and **G** vectors (as shown in figure 1.7) would always create systematic average wind along the path of the circulation integral. Even if the detailed local winds could not be predicted, the overall average wind across a section of coastline could.

fundamental forces of an ethereal view of nature but in explaining the machinations of one of the most visible forces—that of the weather.

Stormy Waters

A number of unusually severe storms caused havoc in Swedish coastal waters in late 1902 and early 1903, and, not for the first time in history, there was a call for some form of storm warning system to be devised. Ekholm seized the opportunity and campaigned for upper-air measurements to be taken by balloons and kites. He planned to use the information gathered from these observations to devise some empirical rules for the forewarning of storms. We can imagine Bjerknes almost at the point of slapping his hand against his forehead in bewilderment at the lack of any underpinning physical theory for Ekholm's ideas.

So Bjerknes set about devising a "rational approach to weather forecasting" based on the laws of physics. In 1904 Bjerknes published a paper entitled "The Problem of Weather Forecasting as a Problem in Mechanics and Physics." This paper was the culmination of a concerted effort to rationalize the methods of forecasting from rules based on experience to ones based on the laws of physics. By this time, although he had not altogether given up hope of pursuing the physics of the ether, he was recognized as an authority on the application of ideas from the physics of fluids to solving problems of flows in the atmosphere and oceans.

Bjerknes explained the physics behind the problem of forecasting the weather in yet another lecture to the Stockholm Physics Society, which quickly attracted attention in the newspapers. Bjerknes developed this lecture into a series of popular articles on the problem. However, we should be wary of attributing Bjerknes's desire to forecast the weather to any pure sense of altruism: fascination with the fact that the problem was amenable to solution via the laws of physics, and in particular those of mechanics, was probably Bjerknes's real motivation. He was also influenced by Hertz's view that prediction was the highest ideal of science, and foretelling the future weather was an obvious goal to be pursued; the world would be explained by science.

The 1904 paper has become a classic, and it has two remarkable features. The first, which we might expect of a classic, is that more than one hundred years later, a meteorologist who reads this paper will instantly recognize the ideas presented by the forty-two-year-old Bjerknes as those that lie behind modern forecasting. Today, meteorologists and climatologists have supercomputers, satellites, and high-tech graphics at their disposal, but the principles underlying the use of these innovations—which had not even been dreamed of in Bjerknes's day—have their foundations buried deep in the seminal ideas of that 1904 paper. The second remarkable feature of the paper, whose subject is one of the most ancient scientific fields of study, is that it is devoid of an introduction or virtually any other reference to those who had worked on this problem before. Aristotle wrote an entire book on meteorology, which stood the test of time for more than 1,500 years. The French philosopher René Descartes considered meteorology such an acid test for his theories of understanding, observing, and explaining the world around us that he devoted a key chapter to it.

The reason Bjerknes's ideas have endured while Aristotle's have not is one of the main themes of our story. The reason Bjerknes dispensed with any lengthy discourse on the history of weather prediction may be that, at the beginning of the twentieth century, forecasting the weather was still something of an arcane practice. Methods based on plotting observations on charts were learned and passed from one practitioner to another; attitudes were parochial and insular. Bjerknes was acutely aware that forecasters almost eschewed the basic laws of physics, and he was determined to do something about it. It would take a half century and the development of computers before others could realize his vision for global weather forecasting, but there were foundations to be laid.

Despite the abundant evidence of the role meteorology has played in stimulating scientific inquiry over millennia, science gave little back to weather forecasters before the twentieth century. A handful of noteworthy and important contributions to atmospheric science, which utilize the basic laws of physics, are scattered across the timelines of history, but few of these contributions improve our ability to predict tomorrow's weather. In fact, the first concerted efforts toward weather forecasting, as opposed to the advancement of meteorological science per se, were made only fifty years before Bjerknes's undisputed landmark paper. The year 1854 holds special significance for weather forecasters for at least two very good reasons. The first is that the Crimean War, although remembered through the words of Tennyson as an indefensible military debacle, led to the formation of the first national weather services in Europe. This initiative was born out of a catastrophe. On November 13 the French and British fleets suffered many losses and were nearly destroyed by a storm over the Black Sea.

The flotilla was anchored, laying siege to the Russians who were fortified at Sevastopol. The portents of a storm loomed on the horizon, and soon hurricane force winds and torrential rain swept along the Crimean peninsula. Among the casualties were a medical supply ship (the *Prince*) with seven thousand tons of aid for the disease-stricken allied troops along with a one-hundred-gun battleship, *Henri IV* (see figure 1.8), the pride of the French fleet. Reports of the disaster dominated the news headlines, and an impatient public demanded an explanation.

Louis-Napoléon turned to one of the world's most eminent scientists for help. In 1854 the prediction of storms seemed impossible, but, in stark contrast, the astronomers were predicting dates and times of eclipses and the ebbing and flowing of the tides with considerable accuracy, and for many years into the future. The repute of the astronomers

Figure 1.8. The unhappy fate of many ships and individuals during the nineteenth century was destruction during the frequent storms. Here Lebreton paints the "pride of the French Navy," the battleship *Henri IV*, grounded on the Crimean shore in November 1854. The public increasingly clamored for governments to do something about predicting the weather in order to make sea travel safer. *Bulletin of the American Meteorological Society* 61 (1980): 1570–83. © American Meteorological Society. Reprinted with permission.

had ascended to new levels when, in 1846, two mathematicians—John Couch Adams in Britain and Urbain J. J. Le Verrier in France—independently predicted the existence of a new planet solely from the analysis of observations of other known planets; not once did they endure cold nights peering through telescopes themselves.

In the early 1840s a Cambridge professor of astronomy, George Airy, devoted much of his time to studying the orbit of the relatively recently discovered planet Uranus. Careful observations of the planet's orbit were made and compared with what should be expected based on Newton's law of gravitation. It gradually emerged that there was a discrepancy between observation and the predicted path. A controversy as to the source of this inconsistency raged for many years—Airy even questioned the validity of Newton's theory of gravity—but a more plausible explanation for the inexplicable wanderings of Uranus would be the existence of another, hitherto unseen planet, whose gravitational field would affect the orbit of Uranus.

In France, Le Verrier, an accomplished astronomer with an enviable track record had also decided to take up the challenge of finding the unknown cause of Uranus's wanderings. By November 1845 Le Verrier had demonstrated to the Paris Academy of Science that the wanderings could certainly be explained by the presence of another unseen planet and were not a result of any error with Newton's laws or with the observations themselves. Then, seven months later, in the summer of 1846, he had calculated the position of the unknown body, but no one offered to search for it.

Frustrated by the apathy of his French colleagues, Le Verrier wrote to Johann Galle, an astronomer at the Berlin Observatory. In his letter to Galle he wrote, "I am looking for a persistent observer who is prepared to spare a little time to examine an area of sky where there is possibly a new planet to be discovered. I came to this conclusion from our theory of Uranus. . . . It is impossible to account properly for the observations of Uranus, unless the effect of a new, previously unknown, planet is introduced. . . . Direct your telescope to the point on the ecliptic in the constellation of Aquarius, in longitude 326 degrees, and you will find within a degree of that place a new planet." Galle asked the director of the Berlin Observatory for time on the telescope to carry out this investigation. Fortunately, it appears that he asked the director on his birthday, when the director was otherwise engaged, and promptly acquired

Figure 1.9. Urbain J. J. Le Verrier (1811–77) predicted the existence of the planet Neptune using equations and mathematical solutions. Were storms also predictable using mathematics? Le Verrier examined the Black Sea weather data and concluded that storms would be predictable by using observational information transmitted by the new electric telegraph. Lithograph by Auguste Bry. *Source*: Smithsonian Institution Digital Collections website.

the telescope for the evening. He girded himself for a laborious task—pointing the instrument to the area of sky and checking every object against a chart showing the positions of known stars. But within an hour of commencing his search, Galle came across an object that wasn't on the charts. He had to check the following night to see if the object had moved against the background of known "fixed stars"—it had—and the new planet, Neptune, was less than a degree from Le Verrier's predicted position. This was a triumph for Newtonian physics and mathematical calculation: the prediction of the location of a hitherto unseen planet, nearly 4.5 billion kilometers from the Sun. It was a coup for Le Verrier, whose reputation soared.

If astronomers could calculate the existence of planets from observations, then surely, Louis-Napoléon must have thought, meteorologists should be able to calculate the arrival of storms. So Louis-Napoléon turned to the new superstar of astronomy for advice concerning whether the Black Sea storm would have been predictable. Le Verrier, who now had good connections with scientists at many observatories, asked for the weather reports from locations across Europe for the period November 10–16, 1854. He and his colleagues analyzed the data and concluded that the track of the storm could have been anticipated if observations had been relayed to forecasters in time.

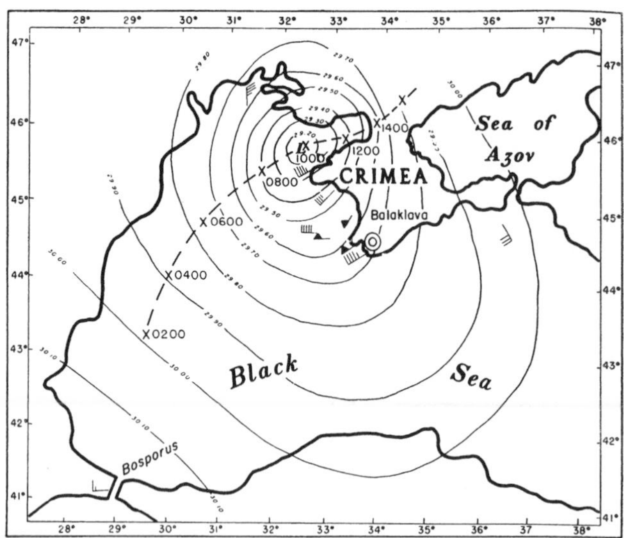

Figure 1.10. The synoptic chart used by Le Verrier to analyze the Black Sea storm. The isobars (curves of constant air pressure at sea level) are shown together with the track of the storm (dashed line from middle left to central upper right). The tightly packed isobars near Crimea indicate the storm force winds there. Their approximately circular shape described the ideal cyclone, or low-pressure system, of the midlatitude regions. Le Verrier suggested that the construction of these pressure maps would allow us to forecast the future movement of such storms. *Bulletin of the American Meteorological Society* 61 (1980): 1570–83. © American Meteorological Society. Reprinted with permission.

The findings of this inquiry led immediately to the call for cooperation in establishing a network of weather stations linked to forecasters by the new technology of the electric telegraph. After measuring at different locations and at regular intervals—air pressure with a barometer, and wind speed and direction with anemometers and weather vanes—the data could be transmitted to forecasting centers by telegraph. The forecasters would then have an up-to-date picture of the weather conditions over a wide area. Each day, the observations of prevailing wind enabled forecasters to predict where the weather systems would then move. The wind speed and air pressure determined the intensity of the conditions. These ideas and methods soon spread throughout Europe and North America, and so began what is perhaps the greatest international cooperation in science: the free exchange of atmospheric data for the purposes of weather forecasting.

The second reason that marks 1854 as having a special place in the history of weather forecasting is the creation of the Meteorological Board of Trade in Britain. The leader of this new department was Vice-Admiral Robert FitzRoy, who subsequently earned the title of the first national weather forecaster. The Board of Trade was created to implement a new agreement reached the previous year by an international conference on maritime meteorology in Brussels. It was agreed at this conference that all meteorological observations aboard naval and merchant vessels should be standardized. Acting on the advice of the Royal Society of London, the civil servants appointed FitzRoy as the chief statistician.

FitzRoy had gained some insight into meteorology from his years in the navy. He had already combined active service with science, having captained HMS *Beagle* in 1831 and selected a young scientist by the name of Charles Darwin to accompany the *Beagle*'s surveying expedition to South America. Indeed, Darwin appears to have spent some time studying his captain, as FitzRoy had a reputation for a quick temper (his nickname was "Hot Coffee"). In 1841 FitzRoy embarked upon a promising career in politics, having been elected a member of Parliament for Durham. But in 1843, on taking up an appointment as colonial governor of New Zealand, his difficult temperament quickly led him into trouble. Within months of taking up his new post, he had fallen out with

Figure 1.11. Robert FitzRoy (1805–65) realized that the storm path that destroyed the new ironclad passenger ship, the *Royal Charter*, with large loss of life in 1859 was predictable, so he set up a storm warning system in the major English ports that saved many sailors' lives, together with their vessels. Photographic copy of a portrait lithograph, made by Herman John Schmidt of Auckland, ca. 1910. *Source*: Alexander Turnbull Library, Wellington, New Zealand.

both the Maoris and white settlers in trying to resolve a dispute over land rights. On being recalled to London, he returned to naval duties and eventually retired from active service in 1850.

During his retirement FitzRoy learned of M. F. Maury's work, which both impressed and inspired him. In 1853 Maury, a lieutenant in the U.S. Navy, published *Physical Geography of the Sea*. This was a descriptive account of currents (especially the Gulf Stream), the variation in salinity, and the depth of the oceans. Mention was even made of atmospheric ozone, systematic observations of which were just beginning to be made at that time, and remarks were also made about the relationship between ozone observations and wind direction.

This work was the culmination of many years spent collating meteorological and oceanographic observations, and Maury put this information to good use: his advice to mariners led to a dramatic reduction in the time for long voyages. In those days, prior to the construction of the Panama Canal, to get from New York to San Francisco, ships traveled via Cape Horn. Using Maury's data, the average duration of this voyage was reduced by 25 percent, from 180 days to about 135 days. Using science paid dividends!

When FitzRoy was asked to take on the new job in the Board of Trade, he threw himself into it with enthusiasm. The establishment of the department and the analysis of data from maritime sources had already led to speculation in the House of Commons that it might lead to the prediction of weather in London twenty-four hours in advance. Following Samuel Morse's demonstration in 1844 of the practicality of the telegraph for long-distance communication, in 1850 James Glaisher in England and Joseph Henry in the United States used the new telegraph to exchange data and construct same-day weather charts in the northeastern United States and in England. Politicians and the public had justifiably high hopes for a new weather warning service.

In 1859 a new fast and robust ironclad ship, the *Royal Charter*, carrying more than 500 passengers and crew and a cargo of gold worth more than £300,000 (at 1859 prices), was wrecked off the Isle of Anglesey. The final toll was 459 lives lost in one of the worst hurricane-force cyclones of the decade. Altogether 133 ships and at least 800 lives were lost around Britain in this storm. FitzRoy was able to gather data on the storm and conclude that its path was predictable. In 1861 FitzRoy began issuing storm warnings based on data gathered from twenty-two observing stations that had been established around the coast of Britain. In 1863 he started forecasting to the public via the newspapers, and this was an instant success. But observations were scarce, so experience and intuition were called for.

Little by little the practice of forecasting evolved into the following typical routine. Every day forecasters would draw-up different types of *synoptic charts* using reports wired to them by telegraph from different locations. The chart to which they paid most attention displayed barometric pressure at ground level. The forecasters would join the places with the same reported pressure readings with *isobars*. This information was then used to draw charts showing the changes in pressure over periods of up to a day. Other variables, notably temperature, precipitation, humidity, and wind, were also plotted on charts. As seen in figure 1.10, the ideal storm cyclone was made up of circular isobars with pressure decreasing to a low in the middle. This surface pressure pattern indicated winds spiraling more and more strongly as the pressure dropped. The forecasters were elaborating the simple rules that both ships' captains and farmers were aware of: that falling barometric pressure had more serious consequences the more quickly it occurred, and so on. The main task was to form a picture of the pressure distribution for the coming day.

The forecasters often assumed that a low-pressure area would continue its observed motion in both direction and speed. But there were no equations (as there were in astronomy) and no calculations, apart from empirical rules. The speed of drift of the cyclone was used to estimate the amount of rainfall with the assumption that the slower the motion, the more precipitation. By the 1860s the Dutch meteorologist Christoph Buys-Ballot had observed that the wind generally blows parallel to the direction of the isobars, and the wind strength is, to a very good approximation, proportional to the rate of change of barometric pressure in the direction perpendicular to the isobars.

Buys-Ballot's empirical law could be used by forecasters in conjunction with charts of the pressure distribution to forecast wind direction and speed. Such information led to forecasts that were usually no more precise than "rainy and windy" or "clear and cold." The wind speed and direction would then blow the weather to the "predicted" neighboring region some hours later. Knowledge of the rules allowed a smart and observant ship's captain to steer away from the worst of a storm.

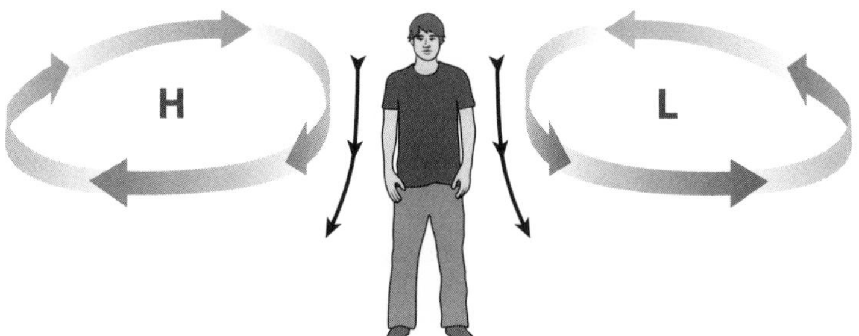

Figure 1.12. Buys-Ballot's law: if someone stands with their back to the wind in the northern hemisphere, then low pressure will be to their left. L is at the center of an idealized cyclone (a large vortex of rotating air several hundred kilometers across and about five kilometers deep), and corresponds to low pressure while the center of the high-pressure anticyclone is denoted H and is to the person's right. In the center, the arrows show the flow of air parallel to the circular isobars. The strength of this central wind is proportional to the pressure difference between H and L divided by the distance between these positions.

As the network of observers grew, larger and larger synoptic charts were plotted. The premise for forecasting was that knowledge of the present weather over a large area could yield a prediction of the future weather at locations nearby. For example, by means of simple extrapolation, a forecaster would predict rain for a certain locality on the basis that the locality lay in the direction of the prevailing wind from the upwind showers.

In addition to their value in forecasting, synoptic charts began to reveal regularities in the weather that had not been observed, or at least recognized, before. In the northern hemisphere it was found that there is generally a counterclockwise flow of air around the center of a region of low pressure (in accordance with Buys-Ballot's law; see figure 1.12), that the wind is usually stronger in the southern part of a low-pressure area, and that temperatures tend to be lower in the western and northern parts of these systems. (In the southern hemisphere these relations are reversed; for example, counterclockwise becomes clockwise.) It was soon discovered that colder weather usually occurs in regions of high pressure (known as anticyclones) during winter. The tracks of cyclones were recorded on charts, and these maps revealed the general tendency in midlatitudes for weather systems to move from west to east. These regularities and patterns increased the skill in issuing forecasts for up to two days ahead.

Despite their limited accuracy, these forecasts were popular with the public, and newspapers printed them on a daily basis. But trouble and controversy loomed: there was a growing divide among the scientific community between those who recognized the utility of the forecasts and those who questioned their scientific basis. The cynics viewed the forecasters with contempt and regarded the use of data and charts as more of an art form rather than a triumph of scientific ingenuity. The skepticism was because the relevant scientific principles embodied in the laws of physics played little or no part in forecasting, which remained almost entirely qualitative, or subjective.

Criticism of FitzRoy's activities gathered support, and in France the situation was little better for Le Verrier: the distinguished scientist was now seen as a crank pursuing a hopeless goal. The drive to establish weather forecasting services had certainly created the stereotypical image of the bungling forecaster. Before 1854, weather forecasting was the domain of soothsayers and sages, of folklore and natural signs such as bird behavior. If the forecast was wrong, then that was just another of life's mysteries. After the efforts of Le Verrier and FitzRoy, who in practice were concerned with the safety of sailors and fishermen, it was perceived that accurate and reliable weather forecasting was proven to be impossible, and those that pursued the ways of the weather became the butt of popular jokes. The level of criticism and ridicule became intolerable for FitzRoy; at 7:45 a.m. on April 30, 1865, he entered his bathroom and cut his throat with a razor.

The Board of Trade and the Royal Society took the opportunity of FitzRoy's suicide to commission a report into his activities. Their findings, published in 1866, were so damning of FitzRoy and his methods (somewhat unfairly, as modern analysis now shows) that forecasts ceased to be issued on December 6 of that year. However, the weather did not relent in its assault on shipping. After considerable losses of life, goods, and vessels followed by strident public demand, the storm warning service introduced by FitzRoy was reinstated in 1867. It took another decade to reestablish the publication of forecasts by the renamed Meteorological Office, but all this time science remained divorced from practical forecasting. So when Bjerknes sat down to write his 1904 paper, he was venturing into territory that the most resourceful civil servants and the most brilliant mathematicians of the nineteenth century had been unable to find, let alone conquer.

Solving the Unsolvable

There is one notable exception among the list of distinguished pioneers of meteorology who did see a clear role for mathematics and physics in weather forecasting. In December 1901, on page 551 of volume 29 of the *Monthly Weather Review* is a paper entitled "The Physical Basis of Long-Range Weather Forecasts," by professor Cleveland Abbe. Over the two decades of 1870–90, Abbe had established a national forecasting agency for the central and east United States, and he had introduced practices that made the agency a world leader. His emphasis on theory to accompany the observational and forecasting procedures eventually led to him being criticized and removed from office, after the weather service was transferred from the U.S. Signal Corps to the Department of Agriculture. From 1893 the Weather Bureau returned to forecasting by empirical rules.

Edmund P. Willis and William H. Hooke wrote in their 2006 paper on Abbe that "Abbe did not view or treat Bjerknes as a competitor, but as precisely the kind of young scholar he hoped to attract to meteorology. Abbe appreciated that Bjerknes' circulation theorem simplified the mathematics in many situations of interest." Abbe welcomed Bjerknes to the United States in 1905 and introduced him to Robert S. Woodward, astronomer, mathematician, and president of the Carnegie Institution; this would have far-reaching consequences, as we discuss in chapter 3.

Figure 1.13. Cleveland Abbe (1838–1916), the first U.S. government chief meteorologist. Initially trained as an astronomer at the University of Michigan, Abbe was an honors student in chemistry and mathematics at the Free Academy (now City College), New York. Abbe's 1879 report introduced present-day standard time to the United States. Reprinted courtesy of the National Oceanic and Atmospheric Administration / Department of Commerce.

Abbe's paper in 1901 contains much greater mathematical detail than Bjerknes's paper that appeared three years later, but Bjerknes's contribution is widely regarded as the start of the road to modern weather forecasting. Perhaps the most important point about Bjerknes's 1904 paper is that it encapsulated a real scientific vision that became a manifesto for how to go about solving a hitherto intractable problem. Bjerknes, spurred on by his great discovery of the circulation theorem and its application to solving practical problems for simple oceanic and atmospheric flows, could now see a way forward.

Bjerknes's plan was stated clearly and firmly at the very beginning of the paper:

> If it is true, as every scientist believes, that subsequent atmospheric states develop from the preceding ones according to physical law, then it is apparent that the necessary and sufficient conditions for the rational solution of the forecasting problem are the following:
>
> 1. A sufficiently accurate knowledge of the state of the atmosphere at the initial time.
> 2. A sufficiently accurate knowledge of the laws according to which one state of the atmosphere develops from another.

Bjerknes introduced the medical terms "diagnostic" and "prognostic" for these two steps. The diagnostic step requires adequate observational data to define the (three-dimensional) structure, or "state," of the atmosphere at a particular time (that is, we describe the "patient's condition"). The second, or prognostic, step requires the assembly of a set of equations for each variable describing the atmosphere, which can then be solved to predict the future states (that is, we make a prognosis of the "patient's future health"). What characterized Bjerknes's approach to forecasting, and what distinguished it from the combination of imagination and folklore that had gone before, was the use of observations combined with the laws of physics—and no more—to predict the changes in the weather. We shall describe these laws in chapter 2.

But Bjerknes realized that the problem was far from straightforward. In the third section of the paper he refers to the inability of the scientists to exactly solve the equations that tell us how three planetary bodies, which influence each other via Newton's relatively simple law of gravity, might continue to orbit each other. The winds in the atmosphere are subject to much more complicated laws, so Bjerknes understood that solving equations exactly for the motion of the entire atmosphere was out of the question. Bjerknes faced a mathematical problem that was, and still remains, theoretically unsolvable.

Bjerknes's key idea to accomplish his goal was to subdivide the problem, which in its entirety is too difficult, into partial problems, each of which is amenable to solution. To this end, he realized that the prognosis need only deal with changes in the weather from one region to the next—say, from one degree of meridian to the next, and over intervals of hours rather than seconds. This is precisely the way in which we go about predicting the weather and climate today using modern computer-generated simulations.

We divide the volume of the atmosphere into a large number of "boxes," as suggested by figure 1.14, where a column of atmosphere is shown layered in the horizontal. Each layer is further divided into smaller boxes, as illustrated by the stipples on the uppermost layer. The

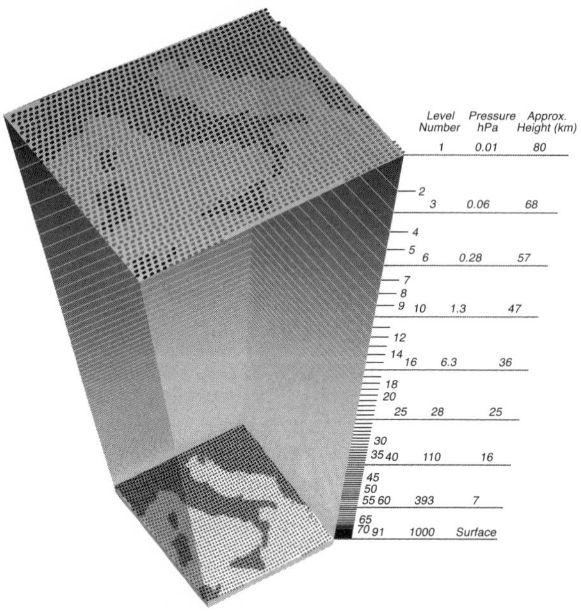

Figure 1.14. The weather boxes, or pixels, used in modern numerical weather prediction models are shown with a column of atmosphere over Italy and the Mediterranean. This structure relates to Bjerknes's original ideas of computing the weather at the intersection of lines of latitude and longitude throughout the atmosphere. The more that computer power increases, the more weather boxes we can use, and the more detailed our forecasts become. © ECMWF. Reprinted with permission.

computer representation of the weather within each little box is uniform, like the colors in each pixel of a digital image. Pursuing this analogy further, we can think of a weather box as a three-dimensional analogue of a pixel, only now the "image" is not a visual one but a mathematical representation of the state of the atmosphere. The numbers represent the weather variables, instead of the standard palette of colors in a computer graphics pixel.

We store seven numbers for each weather pixel: three numbers are used to specify the strength and direction of the wind (two numbers for the speeds in the horizontal directions, and one number for the speed in the vertical direction), three more numbers specify the pressure, temperature, and density of the air, and one more number is used to specify the humidity. This set of numbers determines the uniform state of the atmosphere at each weather pixel so that the entire collection of pixels creates a huge three-dimensional digital representation—or "snapshot"—of the state of the atmosphere at any one moment in time. The problem of weather forecasting is to predict how these weather pixels change from one moment to the next, to create our vision of future weather.

To get a different perspective on these ideas, think about the following idealized picture. Imagine a scene of unspoiled countryside, basking in the warmth of a summer's day. Against the background of a changing sky, mature trees carry their full summer foliage. Across the far side of a meadow, which is a mass of wild flowers and grasses, a group of haymakers is busy at work as darker clouds hint at the possibility of a shower. In the foreground, adjacent to a cottage overlooking a gently flowing stream, two horses stand harnessed to a cart called a hay wain. Of course, this scene has already been captured by human imagination. John Constable's *The Hay Wain* is one of the most famous landscapes of all time.

Invariably, the majority of us gaze on such masterpieces with a sense of wonder, having perhaps never progressed beyond "painting by numbers" ourselves. We find it difficult to comprehend how the artist captures the quintessence of a scene by conveying a sense of motion or activity in a single picture; dynamic skies are a hallmark of Constable's work. Although it appears ridiculous to even imagine trying to mimic Constable's work by a crude "paint by numbers" technique, digitized versions of *The Hay Wain*, which we can find on the Web, really do represent the scene by numbers. These numbers encode the different colors

Figure 1.15. John Constable, *The Hay Wain*, 1821. Image courtesy of Art Resource / The National Gallery, London. Constable's picture is represented in the computer used to produce this book by a large set of numbers, organized in terms of the positions of pixels, just like our computer and TV screens.

and shades in pixels, and if we use a large number of very small pixels, so that the resolution gets fine enough, most of us accept the result.

In the same way that numbers are used to represent colors in a digital image, we represent a snapshot of the weather by using arrays of numbers to specify the values of the variables that represent the state of the atmosphere. Calculating how each weather pixel will change requires powerful supercomputers and large amounts of memory, so limitations on available computer power and memory force a trade-off between the geographical coverage of models, their resolution, and the detail that we can expect from them. The typical horizontal dimensions of a weather pixel used in global weather forecasting out to a week ahead is of the order of a few tens of kilometers, and the pixels are typically several hundred meters deep; horizontal dimensions of about a hundred kilometers are used for seasonal prediction and for simulating climate change over several decades. The horizontal dimensions are reduced to a few kilometers for models of local regions, which provide more detailed forecasts out to two or three days. (See figure CI.3, where the

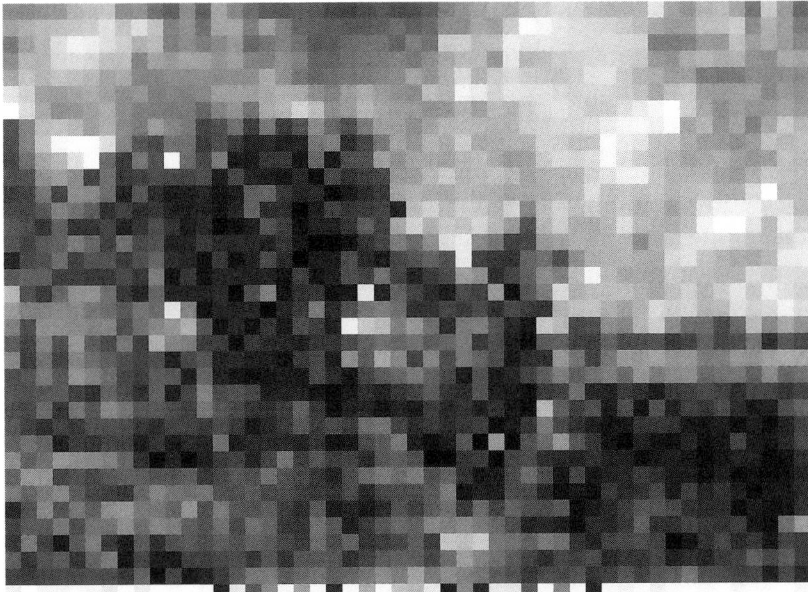

Figure 1.16. An "out of focus" image of *The Hay Wain*, produced using a larger pixel (each pixel is just one shade). We can distinguish between the fields and trees and the sky, and we can just about make out the darker clouds from the blue sky. Just as the size of the pixel in this image limits what we can resolve, so the size of the weather boxes, or pixels, in computational models limits the detail of the weather forecasts—the models today, and for the coming decades, produce forecasts with detail that is closer to the second figure than the first. Bjerknes's idea of calculating the weather at the intersection of degrees of latitude and longitude would correspond to a very "out-of-focus" view of weather.

better pixel resolution leads to much better rainfall description.) Time is the other important dimension; the length of the interval between the "images" is typically anything between one and thirty minutes, depending on the resolution of the model. So, even with state-of-the-art supercomputers, the models represent an out-of-focus and intermittent view of the weather.

Having worked out how to represent a snapshot of the weather at any one moment, our algorithm for calculating the weather forecast then proceeds as follows: on each weather pixel, observations of the current state of the weather are used to compute what the state will be some little time later, according to rules based on the laws of physics. Bjerknes stated the necessary laws of physics, but he did not carry out any calculations in the way we just described. However, he was acutely aware that combining the laws of motion (which govern the winds and pressure patterns) with those of heat and moisture would be a delicate business. He referred to the two phenomena as being joined by a thread—a vital thread that could so easily be broken or tangled.

As figure CI.3 shows, the localized rainstorms, where heat and moisture processes interact most dramatically, must travel with and influence the very winds that are coupled to the pressure contours. Weaving these processes together to make up realistic weather pixel behavior would be at the heart of successful weather prediction by computers.

The whole process may sound somewhat algorithmic, or formal and simple rule-based—and it is; today the calculations are carried out on supercomputers because in any one forecast, many trillions of calculations are involved. If we are to keep pace with the changes in the weather and broadcast the results to users before the weather becomes history, we need to perform this mind-boggling number of computations in only a few minutes.

Further, the trillions of computations all have to "add-up," and the laws of physics tell us how to "balance the books"; it's a bit like a huge exercise in accountancy. However, there is a sting in the tail. Our "weather accountants" may find each new prediction influenced by the calculations as they proceed—the whole problem has the potential for complex and sometimes runaway feedback. Just as controlling audio or electrical feedback is vital for good amplification in concert halls, learning how to control the feedback in atmospheric motion was one of the great challenges to face the pioneers who began implementing Bjerknes's program on computers in the early 1950s.

Bjerknes was aware of many of the difficulties, but he was undaunted in the face of all this and more complexity. He was convinced that the circulation theorem would help forecasters cut a swath through many of the time-consuming calculations, and weave the delicate thread that ties the physics of heat and moisture to the physics that governs the wind and air pressure.

A Ghost in the Machine

Bjerknes's remarkable scheme for a rational approach to forecasting via mathematical computation is precisely the one encapsulated by what we now call his "vision." As a scientific manifesto, it was so attractive that it won support from the Carnegie Institution in Washington, D.C., for a period of thirty-five years commencing in 1906 (a most farsighted and generous grant program that would lead to the foundation of modern oceanography and meteorology). It would be another fifty years before computers became available to do the numerous required calculations in a realistic time, and Bjerknes would just live to see that era. But modern computational methods would also bring to light a problem that threatened to undo Bjerknes's vision—the first appearance of this problem was in the work of one of Bjerknes's contemporaries.

A year before Bjerknes published his disarmingly clear vision of a rational approach to weather forecasting, one of his former tutor-cum-supervisors, Henri Poincaré, a star mathematician of Paris, published an essay entitled "Science and Method." Under this brief, innocuous title, he wrote with the same clarity that Bjerknes would achieve a year later; but unlike Bjerknes, who knew that scientists would share his conviction, Poincaré wrote

> If we knew exactly the laws of nature and the situation of the universe at the initial moment, we could predict exactly the situation of that same universe at each succeeding moment. But even if it were the case that the natural laws no longer held any secret for us, we could still only know the initial situation *approximately*. If that enabled us to predict the succeeding situation with the same approximation, that is all we require, and we should say that the phenomenon had been predicted, that it is governed by laws. But it is not always so; it may happen that *small differences in the initial conditions produce very great ones in the final phenomena.* A small error in the former will produce an enormous error in the latter. Prediction becomes impossible, and we have the fortuitous phenomenon. (emphasis added)

This message was to shake the ground beneath all those who stood on the firm principles of Newtonian physics. His words heralded the beginning of a new era in the age of science. Starting with Galileo and Newton, the earlier era had lasted almost three hundred years. Throughout, there was steadfast belief that the laws of nature hold the key to the predetermined future, known as determinism: the accurate and reliable prediction of all future events, based on our knowledge of the current situation and use of the laws that tell us how conditions will change. In

1903 Poincaré discovered "unpredictability" in practice: the phenomenon we now call chaos.

Poincaré's discovery undermined the common rationalist vision that Newton's laws implied a universe that worked like clockwork and was ultimately predictable, given enough knowledge and effort. His discovery arose from studying the very problem that Bjerknes was to allude to in his paper a year later: the motion of three planetary bodies under the influence of gravity.

As the nineteenth century was drawing to a close, Poincaré realized that two centuries worth of effort by mathematicians to perfect ways of solving the equations of physics was leading to an impasse. His groundbreaking discovery was made while devoting his entire energy to a mathematical problem that became so topical that it featured regularly in polite conversations in select salons of Paris. The problem sought to address whether the solar system would remain "stable" for all time. By "stable" we are asking whether a planet, or planets, might eventually crash into the Sun, or into another planet, or fly off into the depths of space. In 1887 King Oscar II of Sweden and Norway offered a generous prize to the winner of a mathematical problem-solving competition. Anyone who could settle this issue of stability would qualify. The prize, which consisted of 2,500 crowns and a gold medal, was to be awarded on the king's birthday on January 21, 1889.

Figure 1.17. Henri Poincaré (1854–1912), one of the outstanding mathematicians of the late nineteenth century. His work on the stability of the solar system transformed our understanding of Newton's laws of motion in the milieu of the prediction of future events.

The eminent German mathematician Karl Weierstrass, one of the leading mathematicians of the day and renowned for his methodical and painstaking attention to detail, was invited to set four problems. Weierstrass worked in pure mathematics and took great delight in formulating his ideas with rigor and abstraction. So, besides receiving the substantial remuneration (about six month's salary for a professor), the winner would receive the recognition and prestige of cracking a major problem posed by one of the undisputed leading mathematicians of the day. This set the scene for something of a mathematical Olympiad.

Weierstrass had tried his hand at the problem of the stability of the solar system, but without success, so he decided to set this problem as one of the challenges. The problem had been tackled by a number of mathematicians earlier in the nineteenth century, and no one had been able to arrive at any definite conclusions. Still in his early thirties, Poincaré was already well known and had significant achievements to his name. Although he was a busy man, the chance of winning the prize was just something that he could not pass up. As a mathematician, Poincaré was a very different individual compared to Weierstrass. Poincaré was always keen to leap headlong into difficult problems, cutting great swaths through the mathematical jungle without caring so much for detail and precision. But even Poincaré could not leap headlong into Weierstrass's problem. As a first approximation, our solar system was a nine-body problem, the Sun and the then-known eight planets. In practice, however, it was more like a fifty-body problem because of the presence of minor planets and moons of the larger planets, which would have very small influences but should not be ignored.

What is important here is that in any system that is unstable, a small change usually leads to a large change (as anyone who has lost his footing while stepping into a small dinghy, or stepping onto ice, will appreciate). If the solar system were unstable, then weak forces, such as those between the planets (including the minor planets), or impulses that might ensue if the Earth was hit by a large asteroid could eventually cause the entire solar system to change dramatically. Is it indeed possible that one day some of the planets will fly off into space, or collide with each other, or with the Sun?

Poincaré immediately recognized that he would have to make some approximations and simplify the problem. He confined himself to the three-body problem and soon found that it was extremely difficult to provide a general solution—as Newton himself had realized two hundred years earlier. To make progress, Poincaré abandoned the conventional methods that had been developed to solve Newton's laws of motion for the solar system—the same laws that Bjerknes knew were formidable—and devised an entirely new approach. Existing methods were designed to give precise, quantitative information such as the position of the planets thousands of years in the future. Poincaré realized that what he required was more qualitative information—information that would enable him to examine many possible futures.

What was needed was a "top-down approach" whereby he could assess the behavior of a system without having to work out all the detailed solutions, so Poincaré developed techniques for exploring what has become known in physics as the "global picture" of dynamics. This is a method for monitoring the behavior of the orbits of the planets (and many other complicated interacting dynamical systems) over long periods of time—thousands or even millions of years, for example—without having to work out all the details of each and every individual orbit. Poincaré did this by examining what mathematicians call the *flow* of an equation, which describes whole families of solutions.

To indulge fans of *Winnie the Pooh*, and to illustrate the idea of the flow of a differential equation, we describe a game of Poohsticks. Imagine that we are standing on a bridge over a river. Now suppose that we have some sticks with different markings, and we throw them into the river at different points upstream of the bridge and watch them float downstream. If the flow is gentle, then the sticks will float along maintaining their relative positions to a good approximation. If, however, the flow is turbulent or chaotic, the relative positions might well be drastically altered. The point is that we only have to follow the paths of a few well-positioned sticks to deduce something about the nature of the entire flow of fluid. We don't need to work out how every drop of water is moving to deduce that the flow is turbulent. Poincaré developed tools for analyzing solutions of equations in the same sort of way. Instead of having to work out in detail all the possible solutions using calculus, he devised methods for following a few solutions—each one represented by a different colored stick in our example, as we sketch in figure 1.18. By studying relationships between these solutions, he was able to infer what the behavior of the entire system would be.

But even with some new mathematical tools at his disposal, the problem remained daunting. Poincaré took the critical step of making

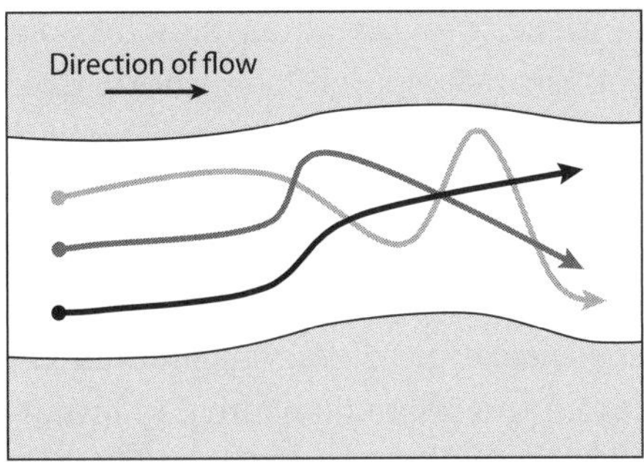

Figure 1.18. Suppose a stream flows under a road, and we stand at one side of the bridge. We do not know what the flow is like under the bridge: it might be smooth or it might be turbulent. By dropping marked sticks into the river at three different locations on one side of the bridge, we mimic different initial conditions in a physical system. The sticks might start by floating along paths that are nearly parallel to one another. However, when the flow of water becomes more turbulent, the trajectories become much more complicated and perhaps even cross one another.

approximations to the problem Weierstrass had set in order to make it tractable. He assumed that the motion of the third body in his system had little effect on the other two. This is a bit like a satellite in orbit in the Earth–Moon system: the satellite has negligible effect on the motions of the Earth and the Moon. Poincaré then searched for solutions of this three-body problem that were *periodic*.

Periodicity means that the state of a system returns to a state it had previously occupied. (A pendulum of a perfect clock is a periodic system: each swing takes the system through the same sequence of states, with each state defined by the position and speed of the pendulum at each moment in time.) By modifying known periodic solutions, he found solutions that were close to the known periodic orbits. He originally assumed that these solutions were periodic too, perhaps over longer timescales, but he did not analyze them further. At the end of the day, he was unable to crack the ultimate problem of the stability of the solar system. Nevertheless, Poincaré made huge inroads into the three-body problem in his entry for the prize: he created the branch of math referred to as "global techniques for dynamical systems." Poincaré submitted a paper that was more than two hundred pages long, and eventually an exhausted panel of judges awarded him the prize by unanimous verdict. Unfortunately, however, a mistake had crept into his calculations, and Poincaré did not spot the error until after the award had been made.

When Poincaré scrutinized the nearly periodic solutions, he noticed that some were not periodic at all. What is more, very small differences between the starting conditions could lead to very different orbits. The orbits could not be calculated with any accuracy in these circumstances. Our Poohsticks analogy is two sticks dropped side by side over the bridge, flowing down the river, and ending up in totally different positions—should we abandon science and our attempts at rational prediction?

Previously it had always been taken for granted that complete knowledge of the state of things, and perfect calculation, would allow total prediction of the future, a vision known by the metaphor "a clockwork universe." But Poincaré's discovery showed that arbitrarily small changes at the beginning of a calculation could lead to radical differences later on. He realized that he had discovered the undoing of Newton's vision of a clockwork predictable universe.

Poincaré opened a radically new scientific window on our universe, and this has even changed our philosophical views. All previous work on Newtonian dynamical systems had been based on using calculus to find particular solutions of the governing equations. Calculus had levered open the natural world to systematic, quantitative investigation and study. Despite the precision of these techniques, their applicability was, and is, severely limited. Although we might apply calculus to solve simple, idealized problems exactly, few problems in the "real" world—from physics to finance—are ever simple. It is a testament to Poincaré's ingenuity that he created techniques that were powerful enough to tackle practical problems. Not only are such problems invariably very complex, but many quantities in the real world are only known approximately, which is something we have to learn to live with. Today we are usually more interested in understanding and predicting the typical behavior of systems that we model using families of solutions.

Was Poincaré's discovery going to undermine Bjerknes's grand vision for a rational approach to calculating the changing patterns of weather—patterns that must surely change according to the deterministic laws of physics? Of course it would. The "rediscovery" of chaos by Edward Lorenz in 1963, while searching for periodic behavior in his simple computer model of the weather, led to the birth of what has almost become an antitheory of weather prediction: the butterfly effect.

The salient issue was epitomized in a remark made in 1972 when Lorenz is reported to have said that the flapping of a butterfly's wings over Brazil might cause a tornado over Texas. Forecasters would have to live with the fact that very small uncertainties in our data of what the weather is like now can lead to very large errors in the calculations of the forecast a few days ahead. Further, it is not only the data that might be inaccurate; our computer models are always approximations limited by the size of the weather pixels. Our knowledge of the relevant physics is also limited. For instance, we do know something about the formation of rain and hail in clouds but certainly not everything. This lack of both accuracy and knowledge means that our attempts at forecasting remain problematic and will always remain so because we will never achieve both perfect accuracy and perfect knowledge.

But to accept the phenomenon of chaos at face value and to conclude that weather forecasters are simply wasting their time is not only defeatist; it also ignores perhaps the most important consequence of Poincaré's work. Mathematicians realized that new techniques would have to be developed to study chaotic systems, and Poincaré's idea of studying families of solutions was the key. The twentieth century witnessed what is perhaps the greatest revolution in dynamical systems since the discovery of calculus by Newton and Leibnitz: the rise of global techniques. These methods reveal the general qualitative behavior of solutions to equations without having to work out the quantitative details of all such possible solutions (which may be numerous and chaotic). These techniques are precisely the ones that forecasters need to deal with the phenomenon of chaos in the weather. If the forecaster can see that a group of forecasts, which all start from slightly different initial conditions, lead to very similar situations—which corresponds to the relative positions of our sticks on the other side of the bridge being roughly the same as when they were dropped in—then our forecaster can be reasonably confident that any one forecast will be reliable. However, if the outcomes from our group of forecasts are very different, then this sends a warning that any one forecast, even if it starts from what we might consider as our best estimate of the current state of the atmosphere, might be totally unreliable (see figure 1.19).

The twentieth century was a formative time for the development of meteorology and weather forecasting; it was a time of serendipitous discovery. Bjerknes's ideas on weather evolved from his father's preoccupation with the ether. Equally, Poincaré's work on the three-body system—considered at the time to be the canonical problem of

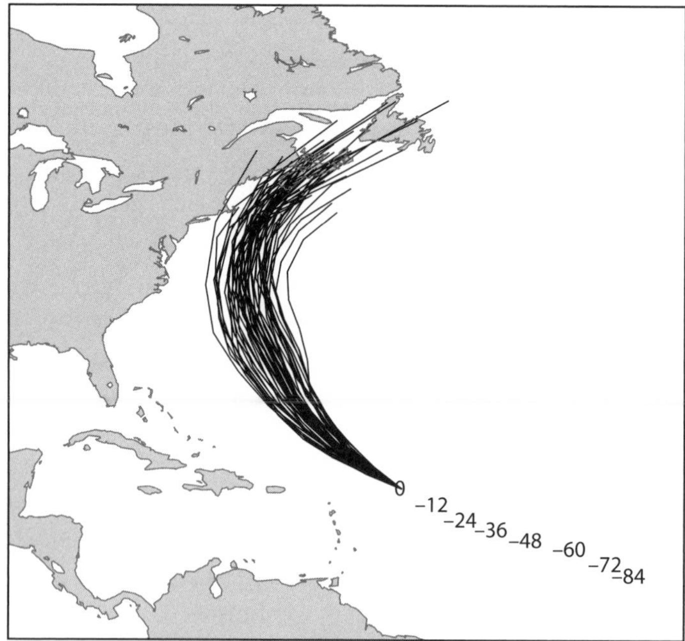

Figure 1.19. Forecasts of the possible track of Hurricane Bill in August 2009. Forecasters need to know how reliable their predictions of events such as the landfall of a hurricane are. By making multiple forecasts, each one starting from a slightly different initial state (or "diagnosis" to use Bjerknes's terminology), and looking at the extent to which the forecasts differ after a period of time, weather forecasters are able to assess how reliable any one forecast will be. This figure shows multiple forecasts of the track of Hurricane Bill, and the numbers indicate its actual position twelve hours, twenty-four hours, etc. earlier. Should a storm warning be issued for the northeastern part of the United States and eastern Canada? This is a "real life" version of the game of Poohsticks we described in figure 1.18. Here we say that the motion of Bill is reasonably predictable over this period. © ECMWF. Reprinted with permission.

deterministic Newtonian mechanics—proved decisive in the development of a radically new view of Newtonian physics. Lorenz's search for recurring behavior in the weather led to computer simulations that demonstrate Poincaré's ideas on the chance phenomenon so vividly.

Bjerknes's circulation theorem weaves the delicate thread that symbolizes the many feedbacks between winds, storms, and the physics of heat and moisture. Lorenz's portrayal of chaos is a constant reminder of the ghost in the machine of weather and climate prediction. The Earth system—the atmosphere and oceans, the icecaps and glaciers, the soil and vegetation, and the animals and insects—is a complex, interacting system with both strong and delicate feedbacks that govern the climate and the habitability of our planet. Our story explains how and why mathematics enables us to quantify complex and subtle feedbacks in weather and climate, thereby providing us with a rational basis for forecasting and prediction, even in the presence of chaos.

TWO

From Lore to Laws

In chapter 1 we introduced the weather pixel, our fundamental unit in building a picture of the atmosphere. Next we need rules to advance the pixels so that we can predict the picture for tomorrow. By 1700 Newton's mathematics was seen to be astonishingly successful at predicting how planets move around the solar system. It would take the next two centuries to extend this mathematics to the movement of the atmosphere. This chapter introduces the basic rules that determine how our weather pixel variables interact with each other to create tomorrow's weather. More sophisticated rules—such as the circulation theorem—emerge a little later in our story.

Renaissance

On January 8, 1913, Professor Vilhelm Bjerknes strode onto the podium of the Great Hall at the University of Leipzig to deliver his inaugural address, which marked the beginning of his tenure of the new chair in geophysics. In his opening remarks, he made the obvious point that, because his position was new, he could not acknowledge the work of his predecessors, as is customary on such occasions. He did, however, take the opportunity provided by the lecture to acknowledge the work that formed the basis for his manifesto for weather forecasting—a synthesis of more than two millennia of scientific endeavor. His 1904 paper was devoid of an introduction; now he had the opportunity to redress that omission and provide the background to his work. The transcript of the lecture, published in the *Monthly Weather Review* in 1914, is another classic paper of our subject.

Bjerknes began his lecture with a provocative yet accurate assessment of the place of meteorology in the natural sciences in the early twentieth century. In the third paragraph he states, "physics ranks amongst the so-called exact sciences, while one may be tempted to cite meteorology as an example of a radically inexact science." He expressed his conviction that the acid test of a science is its utility in making predictions, and drew a stark contrast between the methods of meteorology and those of astronomy. Bjerknes noted that meteorology had continued to be dominated by a form of philosophy rather than by "hard science." Forecasting the weather had remained the province of sages and soothsayers. Sayings such as "Red sky at night, shepherd's delight; red sky in the morning, shepherd's warning" might lack scientific explanation but were useful in certain regions of the world. Although it is easy to scoff at such practices, our ancestors had to produce sufficient food to survive, and weather lore was an important subject. In 1557 an English farmer, Thomas Tusser, published the *Hundreth Good Pointes of Husbandrie*, in which the year is divided into months and all the duties to be carried out on the farm, in the garden, and in the house, are set out with references to how the weather influences what we need to do: "And Aprill his stormes, to be good to be tolde; As May with his flowers, geue ladies their lust; And June after blooming, set carnels so just."

But Bjerknes planned to change all this. His program to make meteorology into an exact physics of the atmosphere involved problems of redoubtable complexity. Using the laws of physics to produce a weather forecast was much more ambitious and demanding than using the laws of physics to predict the return of a comet. Bjerknes also realized that the practical value of such an exercise would be worthless because a rational forecast would take so much longer to calculate than the weather would take to change.

Despite the complete lack of the technology required to produce a timely forecast, Bjerknes concluded that if the calculations agreed with the facts, then scientific methods would be vindicated and the victory would be won. Meteorology would then become an exact science, a true physics of the atmosphere. He concluded, "The problem of accurate precalculation that was solved for astronomy centuries ago, must now be attacked in all earnest for meteorology."

It is perhaps not surprising that Bjerknes aspired to follow in the footsteps of the astronomers. More than half a century later, the scientific world still acclaimed the work of Adams and Le Verrier for their independent calculations that predicted the existence of Neptune. The fact that the predictions were the result of careful calculations based on Newton's laws alone was proof of the power behind the science. This utility was evident in everyday life too: the *Nautical Almanac* published the times of sunrise and sunset, the phases of the Moon, and the times of high and low tides, and these events were calculated using physical laws. Put all this together and it is easy to understand why scientists hoped that by observing a change in the wind we would one day be able to predict, via pencil and paper, the approach of a storm. In fact, some believed that a "weather almanac" would appear before too long. By 1912 both airships (as shown in figure 2.1) and biplanes were becoming more widely used, and these modes of transport were just as vulnerable to the weather as shipping. The loss of life from storms was well known. Now unknown and unusual higher altitude weather threatened aircraft safety as well.

Physics had been loitering at the periphery of meteorology for thousands of years but had made few inroads. Our distant ancestors tabulated and occasionally recorded in some detail the observations of many different types of weather phenomena. The ancient Greeks laid important foundations for modern physics, and meteorology was close

Figure 2.1. The first airship of the U.S. Signal Corps at Fort Myers in 1908 was mainly used for observation of troop movement, gun placement, and such, before and during battle. Buoyancy keeps the airship floating aloft, alongside, and sometimes within, the clouds. Reprinted courtesy of the National Museum of the United States Air Force.

to the heart of their science. The Greek philosopher Aristotle wrote *Meteorologica*, meaning "a study of things that fall from the sky," in about 340 BC. This was the first-ever comprehensive book on meteorology; it defined the subject and was a source of inspiration that led to a development of meteorological thinking. The treatise included theories about most of the common manifestations of weather: wind, rain, clouds, hail, thunder, lightning, and hurricanes. His explanations were founded more on a philosophy of life and were based on the elements of earth, fire, water, and air combined with rational discussion, rather than the solid foundations of an experimentally based physics where everything was measured. In fact, many of his explanations were wrong, but this did not prevent his four-volume text from being considered the most authoritative work on the subject for almost two thousand years. This is because much of the book contains, for instance, amazingly accurate descriptions of the winds and types of weather they bring; this was a serious attempt at the practical forecasting that we acknowledge today. To someone like Bjerknes, Aristotle was credited with creating meteorology as a branch of science, albeit an inexact branch.

Archimedes of Syracuse was a great scholar of this ancient Greek tradition who discovered fundamental laws of physics, together with some ground-breaking mathematics. In contrast to Aristotle, his contributions should certainly be classified under the heading of "exact science." Archimedes's ideas were based on a far greater insight into the role of mathematics in creating equation-based laws of physics, together with his respect for experimental testing of each law. Here "laws" mean mathematical rules relating quantities, and "experimental testing" means measurements that lead to quantitative evaluations of the variables governed by the laws. So proofs and explanations based on words are replaced by rules that are tested with numbers. This new type of scientific description translates readily to computers.

Archimedes is, of course, famous for his legendary shout of "eureka!" on discovering the principle of buoyancy while taking a bath. He realized that if the object is heavier than the amount of water displaced, it will sink, while if it is lighter, it will be buoyant and float. In short, Archimedes discovered the basic principles of *hydrostatics*—the science of fluids at rest. The principle of buoyancy describes how less dense fluids rise above denser fluids (such as oil floating on water), and, nearly two thousand years later, it provided the first clues to answering some

Figure 2.2. Archimedes of Syracuse (287–212 BC) said that given the appropriate lever and pivot point, he could move the world. The Archimedes screw remained the most effective way to lift water, especially for irrigation, for more than two thousand years. But we especially note here his discovery of buoyancy, a principle we see in action every time clouds float above us.

basic questions about the way air circulated around the Earth. In fact, buoyancy is one of the key ingredients in explaining the mists and cloud patterns we see on a daily basis in many places around our planet.

For nearly two thousand years, Western theoretical knowledge of weather phenomena was dominated by the writings of the ancient Greeks. It was not until the great rebirth of Western learning in the sixteenth century that things began to change decisively. The developments in both the Middle East and the Indian subcontinent over the preceding centuries in arithmetic and algebra helped lay the foundations for the changes. The Renaissance was the beginning of the end for the mysterious world in which the universe was governed by magic rather than by rational laws based on careful experimentation. The world of earth, water, air, and fire, with physical phenomena arising from their judicious mixing, was being rethought as experiments and newly invented instruments allowed more accurate measurement of length, time, weight, speed, and so on.

It was a period of great discovery and invention that marked the beginning of a revolution that would usher in modern science, engineering, and technology. Leonardo da Vinci, perhaps the most brilliant artist and engineer of this period, had a clear vision of what constituted good and acceptable scientific practice. He wrote, "It seems to me that any science which is not rooted in experiment, the father of all truth, is empty and full of errors." Nearly a hundred years later Francis Bacon, a pioneer of the new scientific approach to understanding the physical world, really stirred emotions when he publicly claimed that too many so-called intellectuals produced theories that neglected nature, eschewed experimentation, and bowed down before the authorities and dogmas of faith alone.

One man who typified both the end of the era of speculation and the birth of modern meteorology was one of the first modern philosophers, René Descartes. In 1637 Descartes published his renowned book *Discours de la Méthode*, in which he expounded his philosophy of true scientific method. The basis of his method was fourfold; translated, he wrote:

1. Never to accept anything as true unless one clearly knew it to be such.
2. To divide every difficult problem into small parts, and to solve the problem by attacking these parts.
3. To always proceed from simple to the complex, seeking everywhere relationships.
4. To be as complete and thorough as possible in scientific investigations allowing no prejudice in judgement.

The second and third principles of Descartes's doctrine contain the essence of reductionism: the process of breaking problems down into constituent parts with the goal of making each part relatively easy to understand, and then combining the parts to create a wide variety of more complex phenomena, which we ultimately want to study, understand, and make predictions about.

Bjerknes clearly adopted Descartes' reductionist principles in his 1904 paper, not only through his advocacy that a few basic laws enable us to predict the weather but also through his suggestion of a method of forecasting based on relatively simple calculations at every intersection of latitude and longitude. We might think of reductionism as a "bottom-up" approach to physics—calculating each and every process or interaction of the constituent parts of a system, such as our atmosphere. In contrast, an application of Bjerknes's circulation theorem is an example of a more holistic approach—holism as the opposite of reductionism—in which a mechanism emerges to describe how several processes are constrained to interact in very special ways. Instead of working out how each individual process evolves and interacts with others, the circulation theorem combines the individual laws for temperature, pressure, density, and wind so that changes in temperature and pressure, for example, must produce certain changes in wind. The result is a "top-down" view of complex physical phenomena such as cyclones, hurricanes, and ocean currents. These are subtle and important issues that we return to later in our story.

Returning to our historical perspective, perhaps the most important point that distinguishes the meteorology of the ancient cultures from that of the Renaissance is that the ancients did not attempt to use physics to explain the causes of different types of weather. The laws of physics, the surest foundations of which were laid during the Renaissance, were to surpass Aristotle's philosophy of weather. Many laws were deduced following a "technology push" in the seventeenth century. At the head of this stood Galileo Galilei, who lived and worked, together with his student and protégé, Evangelista Torricelli, in Italy. It is to Galileo, working with his students, that we owe credit for the invention in the early 1600s of the first meteorological instruments: the thermometer and the barometer. This provided the means of establishing the essential empirical facts, or truths, behind Descartes's first and fourth principles by finding

Figure 2.3. Galileo Galilei (1564–1642), physicist, astronomer and one of the leaders of the Renaissance, who, together with his protégé Evangelista Torricelli, invented and developed the barometer, the thermometer, and other instruments regularly used by meteorologists to measure and observe weather conditions. Galileo realized that if air was something, then it had weight, and he wanted to know how heavy it was. Portrait by Justus Sustermans painted in 1636.

more reliable observations of the actual behavior of our atmosphere. So began the foundation for the whole science of modern meteorology.

The invention of the barometer is usually ascribed to Torricelli. The instrument consisted of a glass tube about one meter long, sealed at one end, open at the other end, and filled with mercury. The tube is inverted and the open end is immersed in a small container of mercury, as shown in figure 2.4. The mercury settles down in the tube, creating a vacuum at the top, until the pressure of the atmosphere acting on the surface of the mercury in the container exactly balances the mercury column. Hence, the height of the mercury column is directly proportional to atmospheric pressure in a carefully made barometer (the renowned Florentine glassblowers were very skilled at drawing uniform glass tubes).

Galileo designed this instrument to confirm experimentally that the air above us has weight. Previously, according to the teachings of Aristotle, it was thought that air possessed "absolute lightness." The correction of errors such as those in Aristotle's version of physics led to a more accurate understanding of the nature of matter, and to a mass/force-based definition of physics.

The proliferation of instruments enabled scientists to observe and tabulate the variations of pressure, temperature, and rainfall on a daily basis in many different countries. Acute observers, able to measure reliably and more accurately than ever before, used this information to identify relationships that previously had gone entirely unnoticed. For

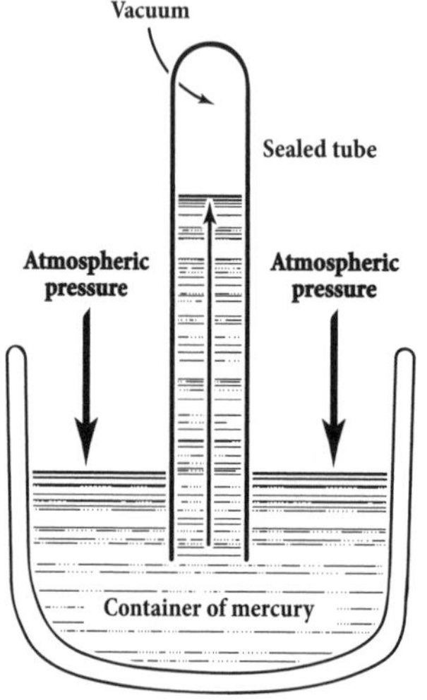

Figure 2.4. Air exerts pressure on everything around it. Torricelli took a narrow straight tube, which was open at one end and closed at the other, filled it with mercury and stood it upright with its open end immersed in a bowl of mercury. A column of mercury about eighty centimeters high remained standing in the tube while above the column surface the remaining part of the tube contained a (nearly) weightless vacuum. This experiment showed that the atmosphere pressed on the surface of the mercury in the bowl to (exactly) balance the weight of the column of mercury.

Figure 2.5. A modern aneroid barometer is shown calibrated for familiar weather conditions. Aneroid barometers, which are the most common in homes today, work on a different principle than Torricelli's mercury barometer (aneroid means "not using liquid"). The main component in an aneroid barometer is a closed, sealed capsule with a partial vacuum and flexible sides. Any change in air pressure alters the thickness of the capsule, and levers that are connected to the capsule then accentuate this change and move the hand.

example, the atmospheric pressure often fell before the onset of rain. Following extensive observations and analysis, barometers were calibrated to indicate the likelihood of showers, rain, and storms on the side of decreasing pressure, and fair, sunny intervals and fine weather along the side of increasing pressure (see figure 2.5).

The collation of observations required organization. In England in 1667, Robert Hooke, instrument maker, mathematician, and City of London surveyor, used his influence to advocate for the coordination and tabulation of meteorological data. He too believed that we would see more regularities or patterns if we had enough observations; therefore, the collation of information was essential. With increased interest in science generally, the seventeenth century witnessed the birth of many societies devoted to furthering the scientific method and bringing together people with ideas and similar interests to share knowledge, both theoretical and practical. The Royal Society of London, founded in 1660, coordinated the recording of meteorological observations, as did the Accademia del Cimento (Academy of Experiment) of Florence and the Académie des Sciences (Academy of Sciences) in Paris.

Mariners on their long sea voyages also enhanced our knowledge of weather by observing the different patterns of wind and rain around the world. Christopher Columbus obtained a detailed knowledge of the trade winds; he used this knowledge to set course for his first voyage to America in 1492, realizing that the strength of those winds in the Atlantic required better mast and sail design to exploit and survive the trades on such long voyages.

About the same time, Vasco da Gama experienced the Indian monsoons, and during the sixteenth century, Spanish and Portuguese sailors completed their first round-the-world voyages and encountered the Gulf Stream. We are well aware of the perils to shipping caused by storms at sea, but just as perilous in the age of sail power were the regions of calms, also known as the doldrums, in which ships would lose the wind and drift helplessly—while the crew ran out of food and water—often for weeks at a time.

So in these two centuries Western thinking gradually changed. Air had weight, so it was definitely "something": what caused the weight? Most importantly, the notion of air pressure emerged, and the measurement of humidity and temperature became possible. The basis of a scientific approach was being argued over, and fundamental principles were being established.

From Comets to Calculus and a Theory of the Trade Winds

By the end of the seventeenth century, a great deal was known about the winds and ocean currents, and the observations of mariners made it possible to create a picture of the typical or average weather patterns at most places around the globe. However, a famous astronomer, not a meteorologist, was the first to use this information to produce a global weather chart. If we take a glance at the history books of science, we find the year 1686 noted for the completion of Newton's book the *Mathematical Principles of Natural Philosophy*, in which Newton expounds his laws of motion and gravitation. We may also note that Daniel Gabriel Fahrenheit, the German physicist and instrument maker whose name is immortalized in the temperature scale, was born. But we would certainly be forgiven for overlooking the publication, on page 153 of volume 16 of the *Philosophical Transactions of the Royal Society of London*, of a paper entitled "An Historical Account of the Trade Winds, and Monsoons, observable in the Seas between and near the Tropicks, with an Attempt to Assign the Phisical Cause of the Said Winds," by E. Halley. This paper is a landmark in the history of meteorology because it heralds the dawn of a new era in which weather patterns are not only systematically observed and tabulated but the laws of physics are also used to explain what causes them.

Edmond Halley is nearly always remembered for the comet that bears his name, but he had many scientific interests and pursuits; indeed this was a hallmark of many of his contemporaries, including Sir Isaac Newton, Sir Christopher Wren, and Robert Hooke, who studied everything from alchemy to architecture. Halley's paper of 1686 is noteworthy for two very good reasons. First, he compiled a meteorological chart—the first of its kind—that depicted the winds over the oceans (see figure 2.6). In particular, he mapped the north and south Atlantic Oceans and the Indian Ocean in some detail, showing the typical or average daily winds.

Second, and this is the crucial point, Halley did not confine himself to simply drawing the map. Halley also attempted to explain the pattern of the winds he observed. One of his motivations for doing this was his observation that the trade winds were present in three separate oceans, and he wanted to know if there was a common cause for them. In fact, natural philosophers had long pondered on what causes the winds. The original explanation of Aristotle nearly two thousand years earlier, that winds were vapors from the earth that rose up into the sky, drawn by the heat of the Sun, was finally being substantiated by arguments involving forces.

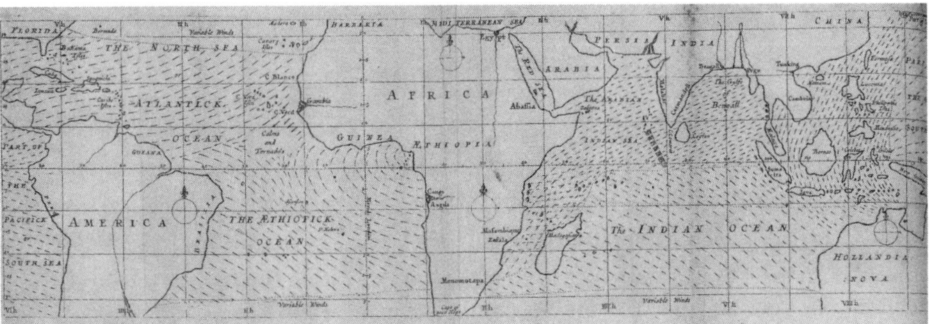

Figure 2.6. The chart from Halley's paper of 1686. The areas of monsoons, trade winds, and the calms are identified. Struck by the large-scale pattern, Halley succeeded in finding a scientific but highly oversimplified explanation. © Royal Society of London. Reprinted with permission.

In offering an explanation for the trade winds, Halley makes use of Archimedes's understanding of buoyancy and of the fact that air expands when heated. In the paper of 1686, he concludes that the Sun is a driving force because, due to the curvature of the Earth's surface, the equatorial regions receive more solar energy per square kilometer than the poles. Halley realized that as the air is warmed by the Sun, it expands and so becomes less dense because there is now less air in a fixed volume. Buoyancy forces caused by the Earth's gravitational pull on the atmosphere increase by an amount that depends on how rarified the warmed air becomes.

A process we now call *convection* transfers heat energy from the warmer air next to the ground to the air possibly a kilometer above the ground, often producing cumulus clouds and thermals, which are the most common signs of this heat movement. On a larger scale, colder, denser, and heavier air from latitudes away from the equator moves toward the equator at low levels in order to replace the warmed air that is rising while the ascending warmer, lighter air eventually moves away from the equator in the upper atmosphere. Halley (on page 167 of his paper) captures this as follows: "But as the cool and dense air, by reason of its greater gravity, presses upon the hot and rarified, 'tis demonstrative that this latter must ascend in a continued stream as fast as it rarifies, and that being ascended, it must disperse itself to preserve the equilibrium; that is, by a contrary current, the upper air must move from those parts where the greatest heat is: so by a kind of circulation, the North-East Trade Wind below, will be attended with a South Westerly above, and the South Easterly with a North West wind above" (the latter occurs in the Southern hemisphere). Halley, in describing the general wind over large parts of the ocean as a current of air driven by the warming effect of the Sun, had started the science of meteorology. However, Halley erred in thinking that the Sun's daily westward path through the sky "dragged" the tropical winds behind it to produce the observed westerly component to those winds.

This average drifting motion is part of what is known as the *general circulation* of the atmosphere, which always tends to move heat from the warm equatorial regions to the colder polar regions. This circulation, where the ground-level air movement shown schematically in figure 2.7 is the atmosphere's response to a steady heating week after week from the Sun, is stronger near the equator and weaker near the poles.

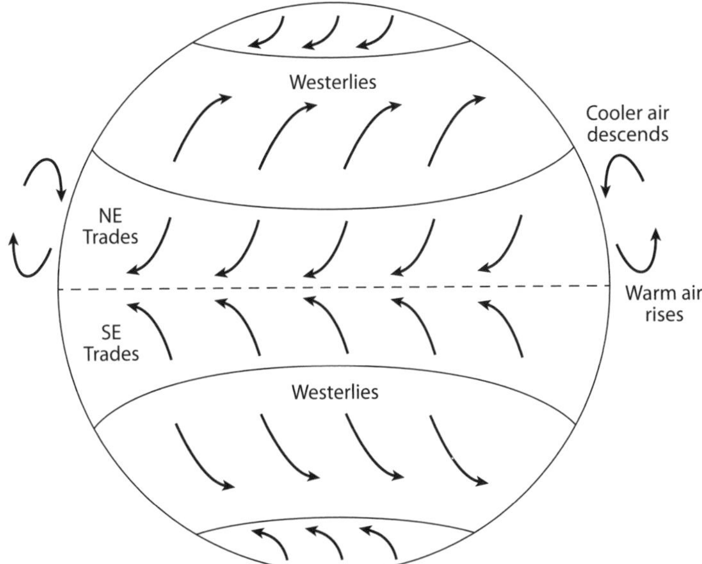

Figure 2.7. The basic pattern of the trade winds and the midlatitude westerlies according to Hadley. The arrows show the average winds at lower, or ground level, around our planet. At high levels in the Hadley Cell we have average winds heading in the reverse (poleward) directions.

The circulation thereby corrects the thermal imbalance caused by hot tropics and cold poles. Thus, our atmosphere is like a huge heat engine moving heat away from the equatorial regions and toward the polar icecaps. Somewhat paradoxically, at ground level we mostly see the cooled return flow from the poles toward the equator, which compensates for the less dense flow away from the equator at high altitudes.

Halley had a clear grasp as to why the difference in the heat received from the Sun between the equator and poles would establish a circulation, and he had the right ideas about the Indian monsoon, but he struggled to understand why the winds are deflected to the right of their direction of motion in the northern hemisphere (see figure 2.7). The circulation Halley envisaged is observed in the tropics, and his explanation of it is essentially correct with regard to buoyant heating.

However, this continental-scale circulation is now known as the "Hadley cell" after the eighteenth-century lawyer and philosopher George Hadley. In 1735 Hadley published a brief paper (also in *Philosophical Transactions of the Royal Society*) entitled "Concerning the Cause of the

General Trade-Winds." In this paper Hadley gave an explanation of the circulation that was much closer to the truth. Halley had correctly deduced that air near the ground flows toward the equator from the north and south, but Hadley began to comprehend the consequences of the Earth's rotation on this flow of air.

Hadley was the first to realize just how the spin of the Earth has a dramatic effect on weather patterns around the world. Because the distance around lines of constant latitude increases from zero at the pole to about forty thousand kilometers at the equator, a point on the equator has further to travel in twenty-four hours in one complete revolution of the Earth while a point at either pole just rotates. Hadley argued that the air flowing from the subtropics to the tropics would acquire westward speeds of the order of 140 kph. But the average strength of the winds in these tropical regions is observed to be much less. Hadley deduced that land and sea surface effects must reduce the wind speed. Thus, the combined effects of the Earth's rotation and the surface effects explain the trade wind that is observed to blow from the east in the tropics.

There is an immediate consequence to Hadley's argument: if there is a net drag on the Earth's surface to the west caused by the whole of the atmosphere, then eventually that drag will slow the spin of the Earth until our planet finally stops spinning on its axis. Hadley therefore deduced that the surface drag in the areas of the trade winds must be compensated for somewhere else. He suggested that this counterdrag would occur in a belt of prevailing westerlies in middle latitudes. To account for the westerlies, he maintained that the air moving aloft toward the pole will acquire an eastward motion relative to the Earth (because it is moving away from the equator and will therefore be traveling around the Earth at a faster rate than the underlying surface). When this air subsequently sinks nearer to the poles, it first returns as a northwesterly and nearer to the equator as a prevailing northeasterly.

More generally, the Renaissance shifted thinking toward a quantitative understanding of nature. Newton's laws of motion and his theory of gravitation remain the most famous of these advances. Such quantitative theories required the development of new mathematical tools, especially calculus.

Calculus became the basis of the theory of *differential equations*, which we need in order to formulate our weather rules in terms of quantitative mathematics—in contrast to the purely descriptive accounts of

Halley and Hadley. As the name "differential" implies, the equations involve the computation of differences between quantities, and these differences are calculated over intervals of space and time. A differential equation uses the notion of differentials to express the rate of change of one variable with respect to another. For example, the differential might be the variation in temperature over a specified time interval. We then calculate the ratio of the temperature difference and the time interval: as the time interval approaches zero, this ratio tends to the derivative, which is here the rate of change of temperature with time. When we plot a graph of the temperature as a function of time, this rate of change is represented by the slope of the tangent to the graph. This time derivative tells us the rate at which it is getting warmer.

The opposite of differentiation (the above construction of the derivative) is known as *integration*. This is what we do when we add small differences together and work out how much something has changed over a fixed interval. In the earlier example of the rate of change of the temperature, integration of the twenty-four hourly changes in temperature gives us the total change in temperature over a day. When we produce a forecast, we integrate the differential equations on which our model is based over the time interval from now to when we wish to forecast. If we were interested in predicting the next solar eclipse, then we would have to add up all the small changes in positions of the Earth and Moon relative to the Sun from now until the time of the eclipse. If we want to forecast the weather, then we add up all the small changes to pressure, temperature, humidity, air density, and wind velocity (just as Bjerknes advocated) until we get to the forecast time.

The techniques of calculus are absolutely central to modern mathematics and physics. In his inaugural lecture, Bjerknes praised those who contributed to the Renaissance; however, he noted that their theories of atmospheric motion and weather were essentially qualitative. That is, they lacked exactness—they did not use the explicit mathematics of calculus—which he foresaw as essential to quantifying meteorological science. The Renaissance had produced the telescope to facilitate observations and data, and the laws of gravitation as rules to relate that data; as a result the astronomers, using mathematics, furthered their subject in leaps and bounds by making predictions of eclipses and planetary orbits. Galileo and his coworkers had developed the thermometer and barometer, and systematic observation and recording of temperature and pressure in many parts of the world soon followed. The basic ingredients for constructing a quantifiable theory of atmospheric motion were present in Newton's laws of motion when we incorporate the effects of gravity through buoyancy. But the problem at the end of the seventeenth century was that no one, not even Newton, had worked out how to even begin to apply the laws of forces, and the motions they cause, to explain the movement of the atmosphere or the oceans.

Parceling Up Fluids

Until the mid-eighteenth century Newtonian mechanics was applied only to solid objects, such as the famous apple in Newton's orchard and the planets of our solar system. The challenge was to extend Newton's laws to fluids. The problem is that a fluid is a collection of a *very* large number of particles whose individual motions might in principle be calculated in the same way as that of a single object such as a planet—except that this would involve a huge number of equations. And by huge, we mean really huge. For example, the number of air molecules in only one liter of air (at room temperature and pressure) is approximately 6×10^{23} (six followed by twenty-three zeroes). So, to compute the motion of all the individual molecules in a litre jug of air would involve $3 \times 6 \times 10^{23}$ equations (the "3" accounts for the three dimensions in which the particles move), which would be an absolutely impossible task, even with the help of today's, or tomorrow's, largest supercomputers. Although, in principle, we could adopt a "bottom-up approach" and apply Newton's laws to each molecule, using calculus to work out its motion, in practice such an approach would be intractable. So, how do we apply Newton's laws to a substance that, by definition, is not a solid and yet contains billions and billions of interacting molecules?

The Swiss mathematician Leonhard Euler was the first to overcome this seemingly impenetrable problem by introducing the idea of what we call a "fluid parcel." A fluid parcel is an idealization of a very small, almost point-like amount of fluid that has two fundamental, if subtle, attributes. First, it is considered large enough to possess many billions of molecules so that it has the crucial physical properties of mass, density, and temperature. And second, a parcel can be small enough so that these average properties do not vary within such a parcel. Fluids are not

Figure 2.8. Leonhard Euler (1707–83) was not only an outstanding mathematician, he was also very productive. He fathered thirteen children, and his collected mathematical works were so extensive that they were still being published decades after his death. Euler earned his degree at the age of fourteen and completed his habilitation thesis at age nineteen from the University of Basel. Euler, disappointed at not getting the vacant professorial chair at age twenty, went to St Petersburg. Even Euler—one of the exceptional mathematicians of all time—wrestled with the intricacies of fluid motion for twenty years before finally publishing his breakthrough in 1757.

composed of physical parcels; indeed, a fluid is a substance that flows freely and conforms to the shape of its container. So even if we were to trace the motion of a "parcel of fluid" (in the usual sense), the parcel would distort after a period of time, perhaps seriously fragmenting and mixing with its neighbors. So a parcel is an idealization but one that can be given a precise mathematical interpretation.

Nearly two thousand years after Archimedes laid the foundations of hydrostatics, Euler, working from 1727 in St. Petersburg with his Swiss compatriot Daniel Bernoulli (who left in 1733), constructed the basis

Figure 2.9. Daniel Bernoulli (1700–1782) came from a long line of brilliant mathematicians. This Swiss family was prone to bitter rivalry, something Daniel was to suffer when he became estranged from his father at the age of thirty. Daniel's book *Hydrodynamica* appeared in 1738—his father Johann's *Hydraulica* appeared about the same time, but he claimed it was dated 1732, to steal his son's credit.

of what is known as *hydrodynamics*—the study of fluid motion using Newtonian physics. Archimedes had worked out that pressure forces are responsible for initiating and maintaining the steady flow of a fluid, and Euler—a virtuoso of the new mathematical calculus—formulated precise equations that tell us how pressure acts to change the motion of a parcel of fluid. Euler deduced that a parcel will accelerate if there is a difference in pressure between the opposite faces of the parcel.

Having worked out how pressure exerts a force on one of his hypothetical fluid parcels, Euler was able to use the result to apply Newton's laws of motion to these parcels. The result was three equations governing the motion of a fluid parcel through the three dimensions of space: one equation for motion in the vertical and one equation for horizontal motion in each of the northward and eastward directions. A fourth equation encapsulates the principle that as a fluid flows through a region, such as a length of pipe, fluid cannot magically disappear. In other words, fluid matter cannot be created or destroyed by motion; we call this the *conservation of mass*.

The resulting four equations of Euler constitute the cornerstone of all ideal fluid mechanics. At the time, Euler was thinking of the flow of water or wine, or even blood—for instance, he and Daniel Bernoulli were the first to suggest ways to measure the flow of blood through our

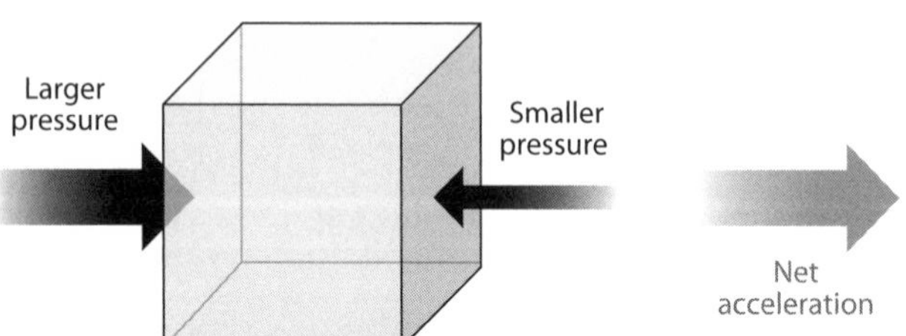

Figure 2.10. According to Euler's equations, pressure forces will accelerate or decelerate a parcel of fluid. Pressure is force per unit area; so if we think of Euler's picture of fluid as a substance composed of lots of parcels, then the pressure forces are created by the parcels "getting in the way" of each other, like people jostling in a crowd. The difference in pressure between the opposite faces of a fluid parcel gives rise to the net force acting on a parcel in that direction. The rate of change of a parcel's momentum (its mass multiplied by its velocity) is proportional to the pressure difference between the opposite faces of that parcel. Consequently, the parcel is accelerated in the direction of decreasing pressure. This rate of change of pressure with distance is known as the pressure gradient.

Tech Box 2.1. Euler's Equations for Fluid Motion

As in tech box 1.1, the notation Df/Dt is used to denote the rate of change with time of any quantity, f, that is defined on fluid parcels. The wind is the movement of air parcels each with velocity, $\mathbf{v} = (u, v, w)$, density, ρ, and mass, $\delta m = \rho\delta V$, where δV is the volume of the parcel. Here the wind has speed u to the east, v to the north, and w vertically upward.

The constancy of the mass of the parcel, expressed as $D(\delta m)/Dt = 0$ (because we do not gain or lose any air mass as the air moves around), means that the density, ρ, of the air parcel is inversely related to its volume. The rate of change in volume is related to the divergence, written as $div\mathbf{v}$, and the rate of change of the air density as the parcel moves can then be written as

$$D\rho/Dt = -\rho\, div\mathbf{v},$$

which is derived by expressing the conservation of mass in the form $D(\rho\delta V)/Dt = 0$.

Newton's law for the rate of change of momentum, $\delta m\mathbf{v}$, is written in terms of the force, $\mathbf{F}$, acting on the parcel as

$$D(\delta m\mathbf{v})/Dt = \mathbf{F}\delta V.$$

Again, using the conservation of mass, this is more usually written as $D\mathbf{v}/Dt = \mathbf{F}/\rho$. Here the force, $\mathbf{F}$, is the sum of the gravitational force, the pressure gradient force (see figure 2.10) exerted by the neighboring parcels pressing on the parcel (usually written as $-\text{grad}\, p$ in terms of the pressure, p), together with any (usually very small) frictional forces.

Euler first wrote down these equations for liquids with constant density and temperature in 1756, where he thought of $\mathbf{F}$ as being the pressure gradient alone. The major complication for the Earth's atmosphere is that we need to study the equation $D\mathbf{v}/Dt = \mathbf{F}/\rho$ when the fluid rotates with the planet, when it is subject to the force of gravity, *and* when it incorporates the effects of heat and moisture, as we discuss in the remainder of this chapter.

veins and to relate it to the idea of pressure. So their ideal fluids were water-like and did not compress as gases do. In formulating these notions mathematically, Euler laid the foundations for modern hydrodynamics in a seminal paper of 1757. We take a closer look at these ideas in chapter 4 because it turns out that, even after 250 years, Euler's equations of motion pose one of the greatest challenges to mathematical physicists.

Matters of State: From Force to Energy

Halley's theory of the basic circulation of the atmosphere was not his only significant contribution to theoretical atmospheric science. He was also the first person to use the laws of physics to work out how pressure varies with altitude as we climb through the Earth's atmosphere. This was an amazing achievement: his research was conducted nearly a century before the first balloon ascents were made in the latter part of the eighteenth century, and the only evidence that air pressure varied with altitude had been provided by intrepid mountaineers. Halley's work gave scientists clues about the structure of the atmosphere way above the tops of the highest mountains that had been conquered. The key formula that Halley used came from an equation that describes how the pressure of a gas varies as the volume in which it is contained is changed. The law of physics that this formula describes is known as Boyle's law. This law was discovered independently by the French physicist Edme Mariotte, and by Boyle's unsung and hard-working assistant, Robert Hooke. They should probably both be given equal credit.

In 1653 Boyle met John Wilkins, the leader of a group known as the Invisible College, who gathered for their meetings at Wadham College, Oxford. This group included some of the most eminent scientists of their time, and they received acclaim when they became the Royal Society of London in 1660. Wilkins encouraged Boyle to join the Invisible College, and Boyle went to Oxford University, where he acquired his own rooms so that he could carry out his scientific experiments.

While in Oxford, Boyle made important contributions to physics and chemistry, but he is best remembered for his gas law. This law appeared in an appendix to his 1662 book *New Experiments Physio-Mechanicall, Touching the Spring and the Air and Its Effects*. The original text, written

Figure 2.11. Robert Boyle (1627–91) was the son of the Earl of Cork, the richest man in Great Britain. Boyle was educated at Eton at a time when the school was becoming fashionable among the wealthy. At the age of twelve he was sent by his father on a tour of Europe, and early in 1642 he happened to be in Florence when, at nearby Arcetri, Galileo died at his villa. Being a staunch Protestant, Boyle had sympathy for the aging Galileo following his harsh treatment (house imprisonment and public recantation of his ideas) at the hands of the Roman Catholic Church. Boyle became a strong supporter of Galileo's philosophy of using physics to understand the universe, and he believed the time had come for a new approach to studying the world through mathematics and mechanics. The Shannon portrait of Boyle, 1689, by Johann Kerseboom. Oil on canvas. Purchased with funds from Eugene Garfield and the Phoebe W. Haas Charitable Trust. Photograph by Will Brown. Reprinted courtesy of the Chemical Heritage Foundation Collections.

in 1660, was the result of three years of work with Hooke, who had designed and built an air pump. Using the pump, they had been able to demonstrate several fundamental physical facts, including that sound does not travel in a vacuum and that a flame requires air to burn. Boyle's law arose from their discovery that, at a particular temperature, the pressure and volume occupied by a gas are related to one another. If the pressure rises (say is doubled) then the volume decreases (that is, halved), and, conversely, if the pressure drops the volume increases as the gas expands.

These findings, when translated into mathematics, lead to what we now call an *equation of state*. The law is written $pV = a\ constant$, where p is pressure and V is volume. The term "state" here refers to the state of the gas, as described by its pressure and density. But temperature turns out to be crucial because it changes the constant in the equation. Before we say more on this temperature effect, we outline the way Halley used arithmetic to solve equations to obtain a most useful formula that lies behind the design of altimeters used by pilots and mountaineers to measure their height above sea level.

In 1685 Halley used Boyle's law to derive the first simple expression for calculating height, or elevation, from pressure. On his graph, shown in figure 2.12, Halley plotted the pressure on the horizontal axis and the volume of a certain mass of air on the vertical axis. This curve turns out to be a hyperbola, as given by Boyle's law pV = constant; see the curve labeled 1, 2, 3, 4 in figure 2.12 with respect to the pressure and volume axes. In math terms, when we climb by the change in height $\Delta z = z_{n+1} - z_n$ from the height z_n at level n to the height z_{n+1} at level $n + 1$, the change in our pressure, $\Delta p = p_{n+1} - p_n$, is due to the change in the weight of the atmosphere above us. This equation $\Delta p = -\rho g \Delta z$ is now known as the *hydrostatic equation*. The hydrostatic equation tells us how the pressure decreases with height. Halley was able to use this equation and Boyle's law to work out a relationship between pressure and altitude, which we describe in tech box 2.2.

The hydrostatic equation is used in many weather forecasting models to this day to represent the fact that, particularly on larger scales, the strength of the wind in the vertical is very much smaller than the strength of the wind in the horizontal. Over much of our planet the horizontal winds exceed twenty kilometers per hour, whereas the average vertical wind is typically less than thirty meters per hour. This is because gravity and the upward pressure gradient force are almost equal and opposite; that is, these forces are almost balanced.

In thunderstorms, the hydrostatic equation may no longer be a reasonable approximation in the vertical because the rate of change of pressure in that direction is no longer balanced by gravity alone: rapid heating and cooling caused by condensation and evaporation will feed energy into and out of the system and generate regions of strong ascent and descent. These regions of updrafts and downdrafts can be very dramatic; severe turbulence is something commercial airline pilots do their best to avoid. (More gentle regions of ascent occur in thermals, which glider pilots and birds use to gain height.)

Halley's work plays an instrumental role in helping us to understand many basic features of "textbook meteorology." Its full potential, however, cannot be realized until we extend Boyle's work so that we take into account variations in temperature, an important consideration

Tech Box 2.2. A Closer Look at the Hydrostatic Equation

The hydrostatic equation tells us how pressure varies with height in the atmosphere. Halley solved it in a way that foreshadows modern computer-based methods. This method calculated the pressure and density in the stratosphere more than 250 years before man actually traveled there.

Halley assumed constant temperature in this air column, so that Boyle's Law applied. Then the change in density $\Delta\rho_n$ causes a corresponding change in pressure Δp_n, where the constant value of the correspondence for the layers of air is R^*. Thus, $\Delta p_n = R^*\Delta\rho_n$. We can use Boyle's Law to eliminate the unknown pressure differences Δp_n, getting $R^*\Delta\rho_n = -\rho_n g\Delta z_n$, which directly relates the change in height to the change in density.

Halley correctly solved this equation to find a formula for H, the height at which the air pressure is p and the density is ρ:

$$gH = R^*\ln(\rho/\rho_0) = R^*\ln(p/p_0).$$

This gives the height, H, above the reference level at pressure p_0 and density ρ_0.

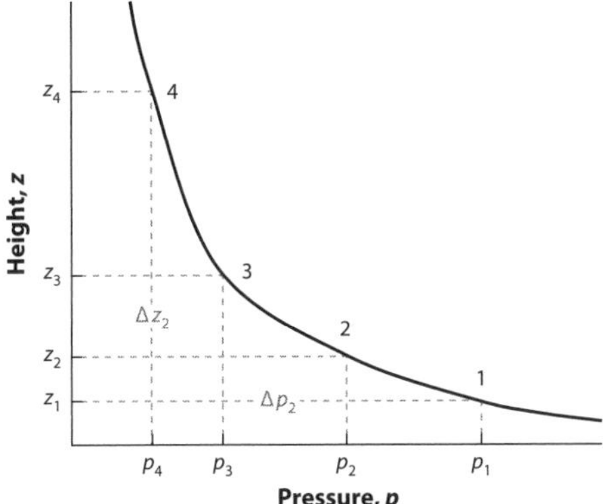

Figure 2.12. We show a given vertical section of atmosphere broken into horizontal layers by the heights z_1, z_2, z_3, z_4. Here $\Delta p_n = p_{n+1} - p_n$ means the decrease in pressure across a layer n of air of thickness $\Delta z_n = z_{n+1} - z_n$. The density in the layer is ρ_n, so that $\rho_n\Delta z_n$ is the mass of air in a vertical column with cross section of unit area within the layer. Then $g\rho_n\Delta z_n$ gives the weight of that air in terms of the strength of gravity g. Halley argued that the increase in pressure as we descend through such a layer is due to its weight, so that $\Delta p_n = -\rho_n g\Delta z_n$.

when trying to work out how pressure varies with altitude. Boyle's law states that at a constant temperature, the volume of a gas is diminished in the same ratio, or amount, as the pressure upon it is increased. Around 1787—more than one hundred years after Boyle first published his results—Jacques Charles discovered what is now known as Charles's law, which states that at fixed pressure the change in the volume of a gas is directly proportional to the change in the temperature. Charles had successfully designed balloons made of varnished papier-mache and filled with heated air. Parisian crowds cheered as the balloons carried people into the skies in December 1783. Charles puzzled over the effects of heating on the air—how hot should air be to raise a balloon by three hundred meters? Somewhat strangely, he did not publish his discovery, and it was not until 1802 that Joseph Gay-Lussac published papers that included this law.

When Charles's law is used in the hydrostatic equation, we get a simple rule for a static (motionless) atmosphere that describes the variation of temperature with height, which is called the *lapse rate*. This quantity is conceptually important for a number of reasons; in particular, it helps us to understand how and where parcels of warm moist air will rise, eventually creating clouds and precipitation. A major lesson is that one of the striking features of weather—rainfall—is triggered by air parcels ascending at rates that are often very small compared with the horizontal winds that accompany the rain. Although wind speed and direction can be measured in the horizontal quite accurately, it is actually very difficult to measure motion in the vertical. So one of the most important quantities in dynamical meteorology—the rate of ascent of air—is one of the hardest things to observe and measure, and even harder to forecast!

We may well ask why it took a hundred years to advance from Boyle's law to Charles's law. There are two main reasons. First, although the behavior of real gases follows Boyle's and Charles's laws fairly well, the laws neglect the effects of moisture and neglect whether the gas is a mixture, for example, of nitrogen, oxygen, and carbon dioxide. Much later, as a result of trying to find a simple model of the behavior of real gases, the concept of an "ideal gas" was introduced in which these two laws are obeyed exactly. Rather like the "perfect fluid" that Euler, Bernoulli, Helmholtz, and Kelvin studied, the ideal gas concept proved very useful. This concept allowed Boyle's and Charles's laws to be combined into what we also call the equation of state, now relating pressure, volume (or density), *and* temperature (see tech box 2.3). This equation, suitably modified to include the effects of moisture, is used in weather forecasting models today.

The second, and major, reason for the hundred-year delay was that scientists had considerable difficulty in working out exactly what the difference was between temperature and heat. Despite the unification of the Boyle/Charles laws into a relationship between pressure, temperature, and density in air, and despite the practical advances being made with the construction of more accurate thermometers, the quest for understanding the fundamental nature of temperature was still ongoing at the beginning of the nineteenth century. One of the deep-seated barriers to progress was finding a precise definition of heat. This may seem a little odd to us, but because scientists at the time were familiar with the idea that heat flowed from hotter to cooler bodies, they thought that heat was a type of "substance" known as "caloric," like water, for example. They also assumed that heat was conserved as it moved from one object to another, like the total amount of fluid, which obeys Euler's equation of continuity.

The Oxford chemist Peter Atkins, in his book *The 2nd Law: Energy, Chaos, and Form,* has pointed out that this view was held by Nicolas Léonard Sadi Carnot, the son of a minister of war under Napoleon and the uncle of a later president of the French Republic. At the time of the struggle between England and France for ascendancy in Europe, Carnot was perceptive enough to realize that industrial muscle would be just as important to major powers as military strength. People were moving from the countryside to the cities as the industrial revolution swept across Europe. At the center of this revolution was the steam engine. Carnot devoted his efforts to analyzing what was needed to improve the efficiency of the steam engine, and in so doing he established a new way of thinking about heat and how we relate heat to mechanical effort, that is, to "work."

In Carnot's description of his ideas, published in 1824, he subscribed to conventional theory that heat was a sort of "ether" that carries warmth. Prior to the invention of the steam engine, the power of flowing water was harnessed to drive machines by means of water mills. Water was a conserved source of power in that the amount of water flowing into a mill was the same as the amount flowing back into the river. Water was not used up in the process. Carnot took the view that the operation of a steam engine was just like the operation of a water mill, so his caloric ran from the boiler to the condenser. Just as the quantity of water remains unchanged as it flows through the mill, so the quantity of caloric remained the same as it did its work. Carnot based his analysis on a conservation law for heat. However, the popularly held view was—not for the first time in the history of physics—incorrect. Heat is *not* conserved in a steam engine. Indeed, heat is not even a substance; it is the outward manifestation of something going on inside materials, whether solid, liquid, or gas. But it would be several decades before it became accepted that matter consists of vibrating molecules, and that warmth is a property of the vibrations.

Establishing the truth about the nature of heat and work began with the experiments by the son of a brewer from Manchester, England. James Joule had the good fortune to be born into a family that was sufficiently well off to enable Joule to preoccupy himself with matters that interested him. By happy chance, the source of the family's wealth was a beer-brewing business that depended on the science of converting one type of liquid into another at precise temperatures.

In the 1840s Joule conducted several careful experiments that confirmed that heat is not conserved in an engine, and he went on to demonstrate that heat and work are mutually exchangeable. That is, work could be converted into heat and vice versa. This led to the concept of the *mechanical equivalence of heat*, and to the conclusion that heat is not a substance but a form of energy. Joule's work did not invalidate the conclusions of Carnot's analysis of the efficiency of a steam engine, but it did correct the reasoning that led to those conclusions. Once again, developing technology allowed better experiments, which validated, corrected, and inspired new scientific ideas.

In 1824 Carnot expounded his theory of ideal heat engines, a study of steam and the efficiency of practical steam engines. He identified the temperature difference between essentially the boiler and the condenser in an engine as the crucial factor in the generation of work; work was generated by the passage of heat from a warmer to a cooler body, and the total energy was conserved in the process. In 1843 Joule discovered this mechanical equivalent of heat in a series of very accurate experiments that explicitly and convincingly enabled the conversion to be calculated. In the most famous experiment, now called the "Joule apparatus" (see figure 2.13), a descending weight attached to a string caused a paddle

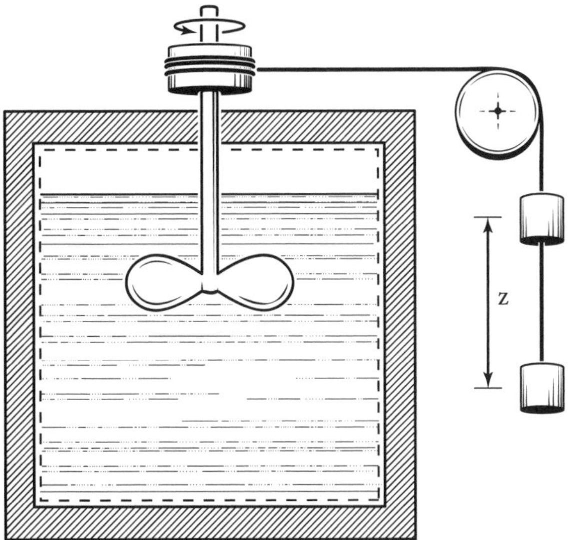

Figure 2.13. The Joule experiment showed that the loss of gravitational potential energy of the mass, M, as it falls by a distance, z, is transformed first into rotary motion energy of the paddle and then into heat energy in the water so that the bath is warmer. The steam engine showed that heat energy could be converted to mechanical energy, so Joule's experiment showed the equivalence of heat energy, related to temperature, with mechanical energy.

immersed in water to rotate. He showed that the gravitational potential energy lost by the weight in descending (over a distance, z, during which the paddles were rotated) was equal to the thermal energy (heat) gained by the water from friction with the paddle. We now measure such energy in Joules to honor his work.

The mechanical nature of heat is not the easiest of concepts to get our heads around, even today. In the middle of the nineteenth century, the concept must have seemed very abstract indeed. Enter once again William Thomson, Lord Kelvin, who heard Joule speak at a meeting of the British Association for the Advancement of Science, held in Oxford in June 1847. From that meeting, Kelvin returned to Scotland with much to contemplate: Joule's refutation of the conservation of heat was shaking the foundations of the existing science. Kelvin also held Carnot's work in very high regard and was concerned that Joule's work would undermine Carnot's conclusions. He therefore set about trying to reconcile their findings.

The basis of Kelvin's approach was to suspect from the outset that there were two laws explaining the conversions between heat and work. About the same time that he set out on his quest to resolve the Carnot/Joule paradox, Rudolf Clausius also saw that the issue might be resolved if there were two principles waiting to be discovered, for then Carnot's conclusions concerning the efficiency of a steam engine might survive despite the findings of Joule. Clausius decided that the notion of the caloric was redundant, and he conjectured that heat could be explained in terms of the behavior of the basic building blocks of matter—atoms and molecules. This heralded the birth of the science of *thermodynamics*, and in 1851 Kelvin published a paper "On the Dynamical Theory of Heat." The key idea behind the work of both Kelvin and Clausius is the notion that it is energy, not heat, that is conserved. Indeed, the emergence of energy as a key unifying concept in mechanics and thermodynamics was a major achievement of nineteenth-century science.

Physics had been the science of force ever since Newton enunciated his laws nearly a century and a half before Kelvin was born. Now energy was about to displace the more direct and explicit concept of force from its hitherto prominent position. By 1851 Kelvin was convinced that physics was really the science of energy. Energy is a word familiar to us today, not least in connection with its industrial and domestic usage, but what exactly does it mean? For now we shall follow our intuition and define it as "the capacity to do work." This definition has an instinctive

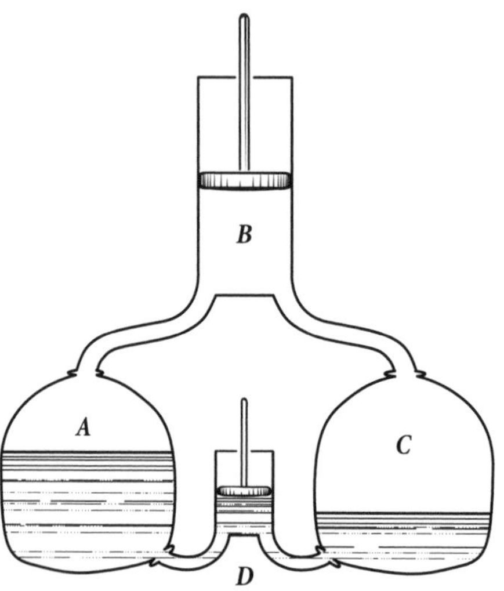

Figure 2.14. The basic anatomy of a steam engine, as understood by the pioneers of thermodynamics. Steam from the boiler A moves to the condenser C via the piston acting in the cylinder B. Water is then returned from C to A by the pump D.

appeal: when we are "full of energy," we have the capacity to do lots of work. Later in the century James Clerk Maxwell and Ludwig Boltzmann would explain these ideas about temperature, work, and energy in terms of the motion energy (kinetic energy) of vibrating gas molecules.

Kelvin and Clausius devised two principles to reconcile the Carnot versus Joule paradox. The first principle is known as the *conservation of energy*, also known as the *first law of thermodynamics*. This conservation law is used in meteorology to describe the movement of heat energy and the related change in pressure, density, and temperature. The second principle is a precise statement that there is a fundamental asymmetry in nature—hot objects cool, but cool objects do not spontaneously become hot. Here is the feature that the two men disentangled from the conservation of energy: although the total quantity of energy must be conserved in any process, the distribution of that energy changes in an irreversible manner. This latter truth about heat energy became known as the *second law of thermodynamics*.

Running concurrently with the Joule versus Carnot debate was the quest to find the last piece of the thermodynamic jigsaw: to devise a universal scale for temperature. In 1724 Fahrenheit published his method of making thermometers and stated that there were three "fixed points" for the calibration: the temperature of a mixture of ice, water, and some salt (believed at the time to be the lowest temperature possible); the temperature of ice and water; and the temperature of the human body. To these points he ascribed the values of 0 degrees, 32 degrees, and 96 degrees, respectively—the basis for the present-day Fahrenheit scale. It is an indication of the crudeness of thermometers in the early 1700s that blood was considered a reliable temperature marker to calibrate thermometers. Now the reverse is true, and we routinely measure daily variation in body temperature to a fraction of a degree.

Carnot had derived a mathematical expression for the efficiency of a heat engine in terms of the temperatures of the boiler and the condenser. Kelvin suggested that a scale of temperature should be chosen so that this expression of work efficiency becomes a universal constant. This means that a temperature-specified amount of heat transferring from a hot body to a cold body would give out the same mechanical effect no matter what the temperature of the hotter body is to begin with. So a system cooling from 100 degrees to 80 degrees would yield the same mechanical effort as one cooling from 20 degrees to a freezing point of water at 0 degrees

Celsius. Kelvin's scale was truly universal in the sense that it was totally independent of the particular physical substance, and the definition of this scale underlines the importance of the notion of energy. Kelvin's scale allowed for supercooling and liquefying helium and many other modern activities that were not dreamt of then—even absolute zero!

Carnot's theory for gauging the efficiency of the steam engine enables us to think once again about the atmosphere as a giant heat engine: the net warmth from the "boiler"—the Sun's rays heating the tropics—flows to the heat sink above the cold polar regions where more heat is radiated away than the sunlight can replenish. As this heat flows from the tropics to the poles, it does work along the way. That is, heat energy is converted into mechanical energy so that the kinetic (or motion) energy of the air parcels increases. This maintains both local weather systems and the overall circulation of the atmosphere around the Earth. The first law of thermodynamics tells us that the total amount of energy involved in this process is conserved; that is, the total amount of heat added first goes to increasing the internal vibration energy of the air molecules—which manifests itself as an increase in the temperature of a parcel of air—and then into performing work on the surrounding air parcels, which keeps the large-scale motion of the atmosphere going.

Combining the equation for the conservation of energy, the equation of state for the gas that makes up the atmosphere, and Euler's four equations of fluid mechanics provides us with six of the seven equations that form the basis of our physical model of the atmosphere and oceans. The final equation describes how moisture is carried around and how it affects temperature. The addition of the seventh equation enables us to describe the Earth-like and life-giving properties of our atmosphere, and as such completes the mathematical model. The condensation of water vapor gives rise to most weather phenomena: cloud, rain, sleet, snow, fog, and dew, for example. The heat from the Sun causes moisture from the ground (or sea surface) to evaporate, and causes heated air near the ground to rise. When rising, moist air often reaches a level at which the temperature is low enough for some of the water vapor to condense, and it becomes visible as cloud (usually cumulus clouds, such as those depicted in the *Hay Wain*). It is truly astonishing to think that the myriad of cloud patterns we see sweeping across our skies can be understood in terms of the seven basic equations in Bjerknes's reductionist scheme. We collect these basic equations in tech box 2.3.

Tech Box 2.3. Math Gets into the Picture: The Seven Equations in Seven Unknowns

The seven equations that form the basis of modern weather prediction are as follows. The first four equations were given in the tech box 2.2 and comprise the wind equations

$$Du/Dt = F1/\rho,\ Dv/Dt = F2/\rho,\ Dw/Dt = F3/\rho,$$

together with the density equation, where $\mathbf{v} = (u, v, w)$:

$$D\rho/Dt = -\rho\ div\mathbf{v}.$$

Here the vector $\mathbf{F} = (F1, F2, F3)$ includes the pressure gradient, the effects of gravity and the Earth's rotation, and frictional forces acting on a fluid parcel. (See the next section for the effect of the Earth's rotation.)

The equation of state says that the atmosphere behaves very like an ideal gas so that the pressure is directly proportional to the product of the density and of the absolute temperature, T:

$$p = \rho RT.$$

Energy conservation, or the first law of thermodynamics, is formulated as follows. The heating (from either the Sun or latent heat of moisture processes associated with cloud and rain formation, for example) both changes the internal energy of the gas, written as proportional to T by means of c_vT, and, by means of the pressure, does work by compressing the gas, of an amount $-(p/\rho^2)D\rho/Dt$, which is written in terms of the rate of change of density as we follow a parcel of fluid (c_v is a constant, known as the specific heat at constant volume).

In summary, we have that the heating rate, Q, on the parcel equals the energy input to the parcel:

$$Q = c_v DT/Dt - (p/\rho^2)D\rho/Dt.$$

Finally we need to monitor the amount of water vapor, q, in the air parcel by means of

$$Dq/Dt = S,$$

where S, the net supply of water to an air parcel due to all condensation and evaporation-related processes, is itself involved in the processes that affect the heating rate, Q, especially where condensation and freezing of water vapor occurs.

In total this gives us seven equations for the seven variables ρ, $\mathbf{v} = (u, v, w)$, p, T, and q.

We might be tempted to conclude our story about the foundation of Bjerknes's vision at this point where we have the seven basic equations. These equations give us the rules for advancing the weather pixels in time, and thus creating a movie that will describe weather evolution. But we must not omit a crucially important ingredient that was included by an American schoolteacher, a man who would justifiably earn the title "the founder of modern meteorology." Euler explains how fluids move in the direction that pressure pushes them, that is, in the direction of decreasing pressure. In contrast, as we discuss in the following section, large-scale atmospheric flow is mostly at *right angles* to the pressure gradient.

The Importance of Spin

Across the centuries, a host of science's leading lights—Archimedes, Galileo, Newton, Euler, Boyle, Charles, Carnot, Clausius, Joule, and Kelvin—laid the foundations, the mathematical laws, upon which modern meteorology would be built. By the middle of the nineteenth century physics had matured to the point where the basic ideas and theories that were needed to formulate a solid scientific understanding of weather and climate were in place. But despite these steady advances, there had been little cross-fertilization of ideas from mainstream physics into meteorology. At the beginning of the 1700s Halley and Hadley explained how physics could be used to understand the trade winds and certain aspects of our climate and weather without solving any equations. The next major step—a step that is now acknowledged as the founding of

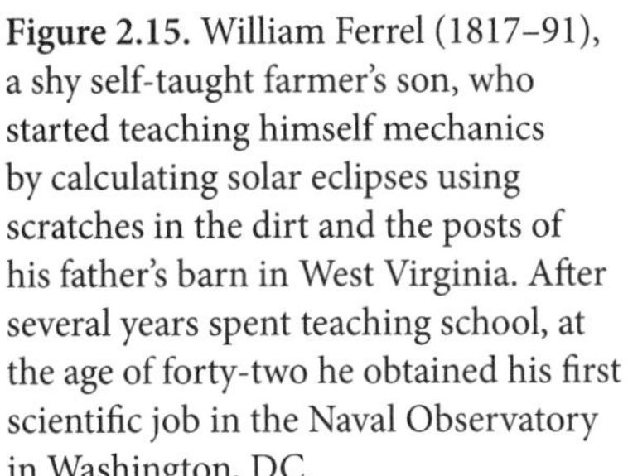

Figure 2.15. William Ferrel (1817–91), a shy self-taught farmer's son, who started teaching himself mechanics by calculating solar eclipses using scratches in the dirt and the posts of his father's barn in West Virginia. After several years spent teaching school, at the age of forty-two he obtained his first scientific job in the Naval Observatory in Washington, DC.

modern meteorology—was taken by one of the greatest "late developers" of all time. William Ferrel earned his place in the history books of science by becoming the first to give a comprehensive analysis of the effect of the Earth's rotation on atmospheric motions using Newton's laws of motion, some thermodynamics, and the language of calculus.

Ferrel was a farmer's son from Pennsylvania. The image of the young Ferrel that has been handed down to us describes a silent, sober, thinking lad: born into a poor family, he made the most of sporadic opportunities to obtain a rudimentary education in the rural environs of West Virginia, where the family had moved in 1829. He had a talent for mathematics but virtually no means of developing his skills because of a total lack of opportunity and resources. In 1832, while working in the fields of his father's farm, he saw an eclipse of the Sun, and this so fired the imagination of the inquisitive fifteen-year-old that he set about learning enough astronomy and mathematics to produce his own tables of solar and lunar eclipses. Ferrel had gone looking for a book on trigonometry to help him with this work but found instead a book on surveying, which turned out to be much more useful. Over the next twelve years he studied avidly, on one occasion riding for more than two days in the winter of 1835–36 to Hagerstown, Maryland, to buy a copy of the book *Geometry*, written by the eminent Scottish mathematician John

Playfair. With his savings and with a little help from his father, Ferrel was able to enroll for a place at college. He later acquired copies of Newton's *Principia* and Laplace's *Mechanique Céleste.* Ferrel could now afford these classic texts because, on graduating from Bethany College, Virginia, in July 1844 at the age of twenty-seven, he had obtained a job as a teacher.

Ferrel published his first scientific paper at the age of thirty-five as an amateur. He was nearly forty when he worked out that the contemporary theories of the atmospheric circulation, which had been developed from Halley and Hadley's ideas, were inconsistent with the laws of mechanics. The missing ingredient in the equations for the purpose of weather forecasting was the effect of the Earth's rotation. Although aware of the Earth's daily spin, other scientists had always dismissed this rotation as insignificant for practical problems. In 1836 the French mathematician Siméon Denis Poisson noted that, due to the Earth's rotation, the trajectory of a shell fired from a gun is deviated to the right in the northern hemisphere. Gaspard-Gustave de Coriolis, director of the prestigious École Polytechnique, derived somewhat earlier the mathematical expression that is incorporated into Newton's second law of motion to account for this effect.

The amount by which a shell from an early nineteenth-century cannon deviates turns out to be small; consequently, what became known as the *Coriolis acceleration* was largely ignored by physicists. (Such corrections became necessary for the battleship armaments a century later.) Coriolis had worked out the equations for predicting the motion of objects relative to the rotating Earth (which arose from his interest in rotating machinery). In 1851 the French physicist Jean Foucault showed that the slow precession, the turning of the plane of swing of an extremely long pendulum (usually more than ten meters in length), is caused by the Earth's rotation, but again this was considered a curiosity. A few years later Ferrel was to use this observation to formulate the definitive equations of meteorology.

The idea that the rotation of the Earth might not only gently affect but actually dominate the behavior of the weather was a profound and acute observation, which is a hallmark of Ferrel's work. The influence of the Coriolis force is what makes large-scale atmosphere and ocean flows so distinctive compared, for instance, to the flow of air around an airplane or a building, or the flow of water around a ship or down a river.

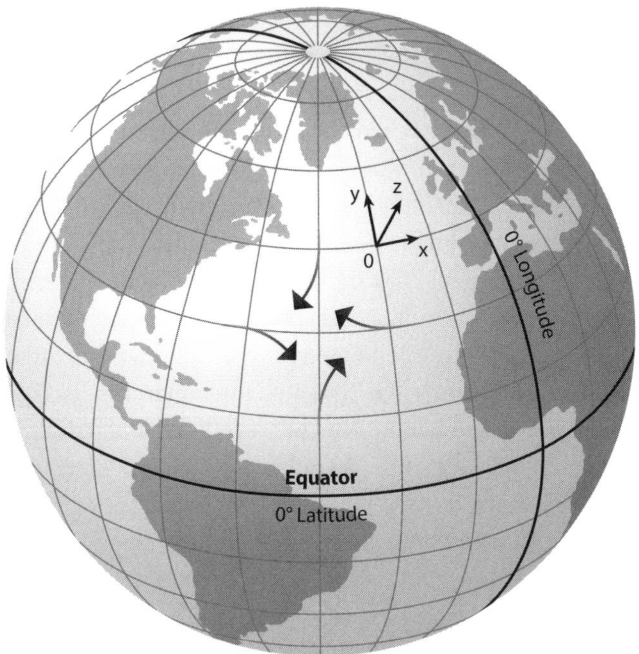

Figure 2.16. The figure illustrates how straight trajectories of wind blowing across the mid North Atlantic, indicated by the broad arrows, are deviated to their right in the northern hemisphere by the Coriolis effect of the Earth's rotation. Eventually Ferrel realized that this caused air to flow mostly along pressure contours rather than directly across them.

Ferrel's approach to meteorology came through first understanding astronomy and the science of mechanics. Ferrel had to teach himself to be accurate when calculating the eclipses of the Sun—an impossible task without careful math calculation. He then spent the years 1858 (when he was forty-one) to 1860 constructing a new theory of the general circulation of the atmosphere around our planet. These investigations were based on an extension to Foucault's ideas, which involved the concept of a deviating force that resulted from the rotation of the Earth. It is clear that he did not know about Coriolis's work, and he used mathematical techniques that were perhaps unnecessarily cumbersome. The lack of clarity in his writing together with the rather curious feature that his admirers were only able to get his first paper on meteorology published in a medical journal were factors that contributed to the initial lack of attention that Ferrel's work received from his contemporaries.

Meteorology in the United States had been driven for nearly fifty years by the powerful but conflicting views of scientists such as William Redfield, who observed the damage trail left by a major storm in New England in 1821 and concluded that all such atmospheric motion was rotary; and James P. Espy, who attacked this view from 1834 to the end of his life. Espy maintained that storms were giant "chimneys" in the sky where the effect of heat and moisture were dominant and the air motion was purely radial (that is, flowing inward at the base of the "chimney" and out at the top). Of course, as Ferrel and Bjerknes later realized, there was truth in both positions. However, Espy would admit to no rotation for the storm winds, and he gave excited and occasionally vitriolic lecture tours of many American states. Espy even encouraged setting fire to the Rocky Mountain forests to create his "chimneys," which would help to bring rain for farms in the Midwest, he claimed.

In contrast, William Redfield quietly embarked on a long study of such storm motions. An engineer from Connecticut, he first observed the tree damage due to the severe storm near his home in September 1821. Then, while traveling to Massachusetts on business, he observed the tree damage of the same storm to be exactly in the reverse direction. Redfield plotted the direction of the felled trees on a map, which he finally published in 1831, showing that the winds of such severe storms, known as cyclones, blew around concentric circles whose center was the center of the actual storm. This work originated the circular theory of cyclones, the "ideal" model of the winter storm of the midlatitudes. Unfortunately, Espy (a lawyer by training), who had used new ideas on the heating of gases and their expansion and convective movement to explain mountain winds (foehns), had a different theory for these cyclones. Espy was employed by the Franklin Institute and later became head of the meteorological bureau of the U.S. Department of War, during which time he produced more than one thousand charts of weather development. Espy commented on rain development and first noted the central steering line of a cyclone but always denied its rotary nature. So Espy followed Euler in claiming that winds were parallel to, but in the opposite direction of, the pressure gradient, grad p, thus overlooking the effects of the Earth's rotation on the large-scale cyclonic winds.

In 1838 the British East India (Trading) Company proposed studying the storms in the Indian Ocean, which affected ships carrying goods back to Europe. From 1839–55 Captain Piddington, curator of the

Calcutta Museum, authorized forty studies on hurricanes in the Bay of Bengal and summarized his results in his first book of advice to sailors.

Redfield studied the major Cuban hurricane of October 1844 and observed a "vortical inclination" of 5–10 degrees of the wind inward from the circular isobars, the contours that showed the pressure levels that ship captains measured. Thus, the horizontal wind was almost perpendicular to the pressure gradient. Spiraling wind motion was observed in reality rather than the circular motion of an ideal cyclone. Redfield had an ally in Lieutenant Colonel Reid of the Royal Engineers, who had witnessed the destructive power of a hurricane in Barbados in 1831. Together their work produced practical guidelines for ships' captains to avoid the worst of a storm at sea—books such as *The American Coastal Pilot* and *The Sailor's Handbook of Storms in All Parts of the World.* Later it was written of Ferrel that his careful study and clear insights, his insistence on logic and reason in place of the war of words, enabled him to resolve, by the 1880s, the long-standing dispute between Espy and Redfield and their various supporters, which had divided American meteorology for nearly half a century.

History has shown us that Ferrel's work marked the beginning of a new branch of science, the subject that has become known as *dynamical meteorology*. Ferrel, in his seventy-two-page paper "The Motions of Fluids and Solids Relative to the Earth's Surface," set out to quantify the qualitative conclusions reached in the earlier paper published in the Nashville *Journal of Medicine and Surgery*. He wrote, "In that [earlier] essay it was attempted to show that the depression of the atmosphere at the poles and the equator, and the accumulation or bulging at the tropics, as indicated by barometric pressure, the gyratory motion of storms from right to left in the northern hemisphere, and the contrary way in the southern, and certain motions of oceanic currents, are necessary consequences of the modifying forces arising from the Earth's rotation on its axis." Ferrel added to the Hadley Cell model of figure 2.7, and figure 2.17 shows his tropical, midlatitudes, and polar zones with their vertical cell rotation. He went on to describe in some detail the importance of rotation, and he concluded that the vanishing of the Coriolis effect at the equator was responsible for the absence of tropical cyclones, or hurricanes, within 10 degrees of it.

Ferrel was able to give a theoretical explanation of Buys-Ballot's law. As we mentioned in chapter 1, in 1857 the Dutch meteorologist

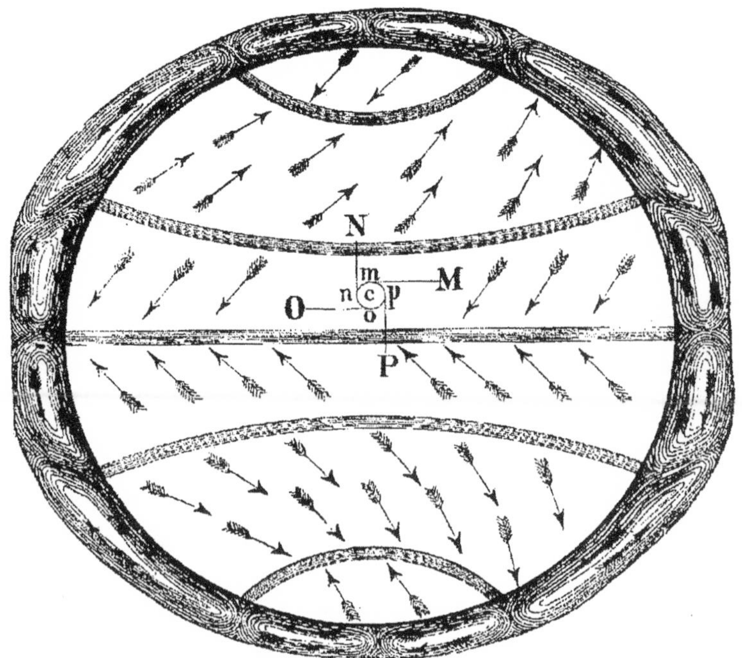

Figure 2.17. One key figure from Ferrel's early paper shows the trade winds near the equator, and the polar easterlies, separated by zones of westerlies. These are the main seasonal winds at ground level. Also shown is the basic mechanism of cyclone rotation. The steadiness (and seasonal variation) of these winds was exploited by ships' captains to cross oceans in voyages often lasting many months. Note that around the edge of the Earth are vertical cross sections of the average winds showing how the air cycles back and returns high in the atmosphere.

Christoph Buys-Ballot formulated an empirical law that the wind direction is generally parallel to the direction of the pressure contours or isobars. Further, the wind speed is roughly proportional to the rate of change of barometric pressure in the direction across the isobars. Ferrel analyzed his equations and showed that Buys-Ballot's empirical law is a consequence of the physical laws of Euler when the Earth's rotation is included. This was a major breakthrough. Surely all such practical laws might also, in principle, be explained one day.

Ferrel was an influential teacher and is deserving of the title "father of modern theoretical meteorology." In 1859, at the age of forty-two, Ferrel was invited to join the staff at the U.S. Naval Observatory, and there he began his proper scientific career after twenty-five years growing up

Tech Box 2.4. Coriolis, the Rotating Earth Term

Consider a location at latitude φ on a sphere that is rotating around its axis through the north pole with angular speed ω. A local coordinate system is set up with the x axis horizontally eastward, the y axis horizontally northward, and the z axis vertically upward, shown in figure 2.16. The local position vector is then $\mathbf{r}_L = (x,y,z)$. The rotation and wind vectors expressed in this local coordinate system (listing components in the order East, North, and Upward) are $\omega(0, \cos\varphi, \sin\varphi)$, and $d\,\mathbf{r}_L/dt = \mathbf{v} = (u, v, w)$. When the point $\boldsymbol{r}_L$ is moving relative to the spinning Earth, its velocity relative to the local fixed coordinates is

$$\mathbf{v}_F = d\,\mathbf{r}_F/dt = d\,\mathbf{r}_L/dt + \omega(0, \cos\varphi, \sin\varphi) \times (x,y,z).$$

When considering atmospheric motion, the vertical velocity, w, is much smaller than the horizontal wind speed; further, $\omega u \cos\varphi$ is much smaller than g, the gravitational acceleration. So when discussing the acceleration terms, only the horizontal components matter. To get the acceleration in the rotating frame we need to differentiate $\mathbf{v}_F$. A complete derivation of this result can be found in university textbooks, such as Gill or Vallis (see bibliography). Then we find (with $w = 0$) the eastward and northward Coriolis-adjusted acceleration terms

$$Du/Dt - f_c\,v = 0,\ Dv/Dt + f_c\,u = 0,$$

where $f_c = 2\omega\sin\varphi$ is called the Coriolis parameter.

In the northern hemisphere, we see that a movement eastward (u is positive) results in an acceleration southward. In general, looking along the direction of horizontal movement, the acceleration is always turned 90 degrees to the right, producing the motion shown in figures 2.16 and 2.17.

The equations for the acceleration of the wind in the eastward and northward directions when there is only a horizontal pressure gradient acting are

$$Du/Dt = f_c\,v - (1/\rho)\partial p/\partial x,$$

$$Dv/Dt = -f_c u - (1/\rho)\partial p/\partial y.$$

The terms with f_c are the additional Coriolis terms that need to be added to **F** in tech box 2.3.

and working on a farm and nearly fifteen years teaching. Ferrel worked at his research for about thirty years before his death in 1891. He had produced approximately three thousand pages of material—work that kept others busy for years afterward. He remained a shy loner all his life, sometimes too nervous even to talk about his own new discoveries. But he had demonstrated from the equations that, on the rotating Earth, air blows mostly along pressure contours, in contrast to the flows studied by Euler and Bernoulli. Ferrel had transformed our understanding of the larger-scale atmospheric flows, and had discovered one of the fundamental forces that control the atmosphere as it moves from day to day.

Although Ferrel derived detailed equations that describe the global weather patterns, and he illustrated the various circulation patterns of our planet's atmosphere in his papers, he was not able to solve explicitly the equations themselves. In deducing how the laws of physics influenced the weather patterns on a day-to-day basis, he—like Halley and Hadley—had to resort to extracting qualitatively useful information without solving the equations for all the detail. What was needed was a person bold enough—bolder even than Bjerknes—to tackle the formidable mathematical obstacles involved in actually solving the equations. And the traditional calculus was going to be of little help, so an alternative would have to be found. In short, two hundred years of scientific innovation had produced a definitive statement of the problem, and the belief that the solutions of the mathematical equations described real weather. Now the problem demanded an actual solution.

In concluding his inaugural lecture, Bjerknes remarked that "under the most favourable conditions it will take the learned gentlemen perhaps three months to calculate the weather that nature will bring about in three hours." Therein lay his concern over the practical value in trying to use the laws of physics to calculate the weather. The first person with the audacity to attempt Bjerknes's challenge head-on shared Bjerknes's view: the real issue was to prove that we *could* calculate a weather forecast, thereby demonstrating that meteorology *is* an exact science. The problem had been diagnosed and the prognosis was clear: if the calculations were feasible, it would spell the end of an era that began at the dawn of civilization. No longer would man have to wonder when to plant, sow, harvest, fish, or hunt; when to set sail, or when to avoid the storms. Physical laws would replace weather lore, and mathematics would deliver the forecasts.

THREE

Advances and Adversity

Lewis Fry Richardson, one of the most enigmatic of British scientists, was the first person to turn Bjerknes's general scheme of diagnosis and prognosis into a precise mathematical algorithm. His calculations were done by hand during the First World War, but the prognosis failed because of a subtle problem in the data from which the forecast was started. Bjerknes's attempts to predict tomorrow's weather developed in a very different way. Electronic computing was still generations in the future, and Bjerknes himself believed that any direct assault on the equations was impractical; so Bjerknes's team developed graphical methods based on charts and the application of the circulation and vorticity theorems to understand midlatitude weather. Their efforts resulted in practical, if qualitative, forecasts.

The Father of Numerical Weather Prediction

News of Bjerknes's grand vision spread far and wide, and it reached the remote Scottish hamlet of Eskdalemuir just before the First World War. Eskdalemuir is hard to locate in an atlas: it lies just north of the border between England and Scotland. Its remoteness made it a good choice for the location of an observatory, so that geomagnetic observations would be least affected by man-made electrical effects (an unusual consideration in the days before widespread commercial and domestic electricity usage). In 1913 this small rural community became home for one of the most extraordinary scientists of the twentieth century, Lewis Fry Richardson.

Richardson had taken a job with the British Meteorological Office and was subsequently posted to the Eskdalemuir observatory as the superintendent. He came from a prosperous northern English family who had made their money from leather tanneries and the corn trade. Richardson's Quaker upbringing meant that he took very seriously the conviction that "science only has to be subordinate to morals." This earnestness and methodical testing of ideas is apparent from a story about him planting money in a flower bed when he was five years old. An elder sister had said that money grows in banks, so Richardson wanted to see if it would grow in the garden too. Having been brought up in the county of Northumberland, Richardson attended a Quaker school in the city of York. His science teacher played an important role in steering him away from business; he spent two years at the Durham College of Science where he studied mathematics, physics, chemistry, botany, and zoology before completing his degree at Cambridge University. His tutor was the world-renowned J. J. Thomson, the Cavendish professor of physics and discoverer of the electron.

After graduating from Cambridge in 1903 with a first class degree in natural sciences, Richardson obtained various employments that utilized his skills in mathematics, physics, and chemistry. One of these was as a chemist with the National Peat Industries from 1906–7, and it was while in this job that he formulated new mathematical methods for solving equations to simulate the flow of water through peat, the soggy brown plant- and soil-based matter that burns well when dried out. By modeling the flow, Richardson hoped to determine the optimal location for drainage channels. The equations were far from easy to solve by any method available at the time, and his methods for solving them were a major departure from the techniques developed by Newton; in fact, as Richardson himself commented some twenty years later, he realized, paradoxically, that his ideas harked back to the sort of mathematics that would have been applied to such problems *before* the invention of calculus.

In 1908 Richardson published a paper explaining how methods based in careful freehand graphical sketches could be used to solve the drainage problem. Two years later he published a paper in the *Philosophical Transactions of the Royal Society of London* that set out his ideas for a more general and innovative method for solving differential equations using arithmetic and algebra (this time applied to the problem of calculating stresses in a masonry dam). This success spurred him on, and

Figure 3.1. Lewis Fry Richardson (1881–1953) volunteered for the ambulance service to work behind the trenches in France during the First World War. During this period he made lengthy calculations and became the first to predict the weather using only numerical methods and the laws of physics. Later he wrote papers on conflict to try to predict, and hence prevent, future wars. His all-round ability and flow of ideas is illustrated by his reaction to the news of the loss of the Titanic, when it collided with an iceberg in 1912. Richardson was holidaying with his wife, Dorothy, on the Isle of Wight on the south coast of England. He asked her to row him in a dinghy in Seagrove Bay, while he used a penny whistle to blow sharp blasts of sound against the pier. Richardson used an open umbrella behind his ear to catch and amplify the echo. The experiment was so successful that he filed a patent in October 1912, with the hope that a practical echo-sounding device would be built to protect shipping during the hours of darkness and in notorious sea mists. © National Portrait Gallery, London. Reprinted by permission.

he applied for a fellowship at King's College, Cambridge. However, his peers there were not impressed by the novel ideas and turned him down. In 1913 at Eskdalemuir Observatory, according to his own account, he first became interested in the problem of solving equations to forecast the weather. In Richardson's imagination, it would be just one step further from his calculations of the flow of water through peat to apply his methods to the flow of air in the atmosphere. He had plenty of time to think about such issues for Eskdalemuir was described as "bleak and humid solitude." This clearly suited Richardson, who once said that solitude was one of his hobbies.

Richardson knew of Bjerknes's manifesto and of his belief in the power of calculation. In the preface to Richardson's book *Weather Prediction by Numerical Process*, published by Cambridge University Press in 1922 after he had completed the groundbreaking calculation we are about to describe, Richardson notes, "The extensive researches of V. Bjerknes and his School are pervaded by the idea of using differential equations for all that they are worth." Richardson was primarily interested in getting answers from mathematics, and his lack of formal training in atmosphere and ocean science enabled him to adopt a fundamentally new approach to the problem of weather prediction. In his book, he follows the great tradition of citing astronomy as a wonderful example of a subject that does indeed get answers from mathematics, and he observes that the *Nautical Almanac*, with its extensive tables of accurate astronomical data, is founded upon the use of differential equations. Richardson considered that to be sufficient justification for a direct assault on the problems of weather prediction using such equations, but he had to find a suitable method for solving them.

At the outbreak of the First World War, Richardson, a Quaker, declared himself as a conscientious objector and therefore was not drafted into military service. This did not stop him, however, from serving his country at this time of crisis. And the story behind Richardson's foray into weather forecasting is set at, of all places, the Western Front. In order to help his fellow countrymen, Richardson, aged thirty-five, resigned from the Meteorological Office and served as a driver for the Friends' Ambulance Unit in the Champagne district of France from September 1916 until the end of the war. Despite serving near the front line, with the effects of inhumanity all around him, he remained resolute and carried to completion a most remarkable calculation. Before he left for the front line, Richardson, following Bjerknes's ideas, thought long and hard about the problem of using physics to calculate a weather forecast. In his 1904 paper Bjerknes alluded to calculating a forecast by calculating the changes in the weather at a finite number of points in the atmosphere—at the intersection of lines of latitude and longitude, for example—but he did not give any specific details for how to accomplish the calculation. Bjerknes appreciated that conventional calculus was a nonstarter as a method for solving the seven equations for the seven unknowns, so in his 1904 paper he advocated reductionism as the logical approach to tackling the problem:

> Everything will depend upon whether we can successfully divide, in a suitable way, the total problem of insurmountable difficulty into a number of partial problems of which none is too difficult.
>
> To accomplish this division into partial problems, we have to draw upon the general principle, which is the basis of the infinitesimal calculus of several variables. For purposes of computation, one can replace the simultaneous variation of several variables with the sequential variation of single variables or groups of variables.

Bjerknes did not develop his ideas further, but the clue as to how to proceed is expressed a little earlier in this same paper: "For to be practical and useful, the solution has to have a readily seen, synoptic form and has to omit the countless details which would appear in every exact solution. The prognosis need only deal, therefore, with averages over sizeable distances and time intervals; for example, from degree of meridian to degree of meridian and from hour to hour, but not from millimeter to millimeter or second to second."

Richardson was bolder than Bjerknes, and his success in obtaining practical answers to practical problems in his earlier work drove him on. In order to put the theory to the test, he devised a method for solving the equations. Euler had formulated the precise mathematical expressions for fluid motion 150 years earlier using calculus to describe the infinitesimal variations of fluid velocity and pressure with location and over time. We recall that a differential represents a tiny fraction of a quantity that approaches but never quite equals zero; it is an abstract concept, an idealization. So Richardson's general idea was to replace this idealization with something much more pragmatic. He realized that to calculate the change in the weather at locations some distance apart, we should replace these infinitesimal differences with finite differences. Richardson then took this idea a crucial step further. By dividing the atmosphere into a large number of three-dimensional pixels, or "boxes," Richardson's inspiration was to find a way of writing the mathematical equations in terms of basic algebra, so that only elementary arithmetic operations—addition, subtraction, multiplication, and division—are needed to solve the equations. Calculus, in the form developed since the time of Newton, was not to be used (see tech box 3.1).

The nature of Richardson's algorithm meant that he was not actually solving the complete problem, but his intuition told him that his method would lead him to an answer that should be close to the solution of the "real" problem. However, Richardson did not rely entirely on intuition and, being methodical, carried out careful tests to make sure his calculations would be accurate enough for the task in hand. It has been said that "ingenious" is literally the correct description, for his methods were based on a deep scientific appreciation of the problem together with a practical engineering approach to solving it.

Working with pencil, paper, slide rule, and a table of logarithms, Richardson computed the first weather forecast. He set out to find the change in the weather over a six-hour period for two locations in a small area of central Europe near Munich. The initial data described the state of the atmosphere over Germany and neighboring countries at 7:00 a.m. on May 20, 1910. Richardson chose the area and the date because unusually good data were available. In what meteorologists today would refer to as a "field campaign," a series of weather balloon observations were collected from coordinated balloon ascents all over Europe. The results had been tabulated and analyzed by none other than Vilhelm Bjerknes, and his analysis of the structure of the atmosphere—from ground level to an altitude of about twelve kilometers—must have been valued by contemporary meteorologists of the day in much the same way that we value the observations of glaciers and polar ice sheets when trying to understand climate change.

We may be forgiven for thinking that computing the six-hour forecast for two points on the globe is hardly an earth-shattering breakthrough, and Richardson's forecast was actually a "hindcast"; he was attempting to predict events that had taken place years before. But even a forecast of this limited scope called for a calculation of daunting complexity. He had to calculate how just more than one thousand variables changed, and to guard against mistakes, he did everything twice! With conviction and determination, Richardson overcame the intellectual and psychological hurdles that must have faced him when he worked out just how many thousands of calculations he would have to perform, and he got on with the daunting task in the appalling conditions of a cold rest billet behind the front line of World War I. With nothing more than a heap of hay and his calculations for comfort, the whole enterprise was to take him more than two years to complete.

On a map of Europe, Richardson drew a checkerboard pattern based on lines of latitude and longitude, which he regarded as partitioning the atmosphere horizontally. The grid he chose had pixels measuring roughly two hundred kilometers on each side (see figure 3.2). He then divided the depth of the atmosphere into five layers so that each column of atmosphere above each pixel at ground level was divided into five boxes (our three-dimensional pixels). The layers were separated by horizontal surfaces at the following elevations: 2.0 km, 4.2 km, 7.2 km, and 11.8 km. Thus, his discrete model divided the volume of the atmosphere being studied into 125 three-dimensional pixels. Time, as well as space, was discrete in Richardson's scheme. He chose a time interval of

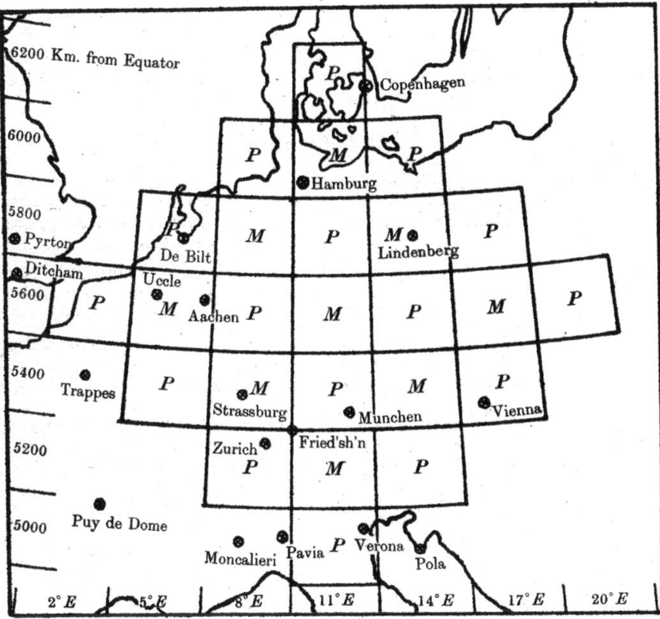

Figure 3.2. The finite-difference grid used by Richardson is based on boxes or pixels that are 200 km by 200 km in horizontal extent—each of these 25 pixels, shown as squares in the diagram, had other boxes stacked on top, so that 125 pixels were used in total. Richardson's scheme was really about ignoring the details and trying to focus on the average overall behavior of the larger weather systems sweeping slowly across Europe. Richardson applied his procedure once—that was enough, for even that involved one-thousand-odd variables and many thousands of calculations. Since the procedure could be applied repeatedly, meteorology might then achieve what astronomy had already achieved: the ability to compute a complete description of the weather at any time in the future without the need to make new observations. The scheme amounts to a precise algorithm (or set of ordered rules) for implementing Bjerknes's ideas from 1904.

Tech Box 3.1. Calculus and the Anatomy of Derivatives

The laws of fluid mechanics and thermodynamics, which we summarized in tech boxes 2.3 and 2.4, were expressed in terms of derivatives that quantify changes in temperature and wind speed, and so on, as we follow air parcels around. But Richardson's calculation is based on fixed pixels, such as the checkerboard of figure 3.2. So we need a formula to convert the parcel-based evolution D/Dt to conventional partial derivatives in which time and space are the independent variables. Euler had already found a way of expressing how changes occurring on or within a fluid parcel can be related to the changes that we would measure at fixed locations in space and time.

Euler expressed his equations in terms of rates of change at particular times and at fixed locations. This is useful in meteorology because, despite the important conceptual idea of a fluid parcel, measurements of the air properties such as pressure and temperature are nearly always taken at a particular location (using satellites and weather stations) and at certain times of the day.

We now explain the basis of Euler's notation, which also provides us with the basis of Richardson's approach. In the middle of the eighteenth century, the French mathematician Jean Le Rond d'Alembert introduced a new concept for the rate of change of a variable that might depend on both location and time. At a fixed location, the change in any variable f over time is written as

$$\partial f/\partial t = (f_{\text{new}} - f_{\text{old}})/\Delta t$$

where Δt = change in time between the new measurement f_{new} and its old value f_{old}. The derivative on the left is defined in terms of the right-hand side when Δt gets smaller and smaller (that is, in the limit when Δt tends to zero). This method also allows calculation of a rate of change as position x is varied to the right or left at a fixed time:

$$\partial f/\partial x = (f_{\text{right}} - f_{\text{left}})/\Delta x,$$

where $\Delta x = x_{\text{right}} - x_{\text{left}}$. We repeatedly use this formula with x measuring distance to the east along the lines of latitude, y measuring distance north along lines of longitude, both at sea level, and z measuring distance vertically upward from mean sea level. Following the practice adopted in most meteorology textbooks, we use a local system of Cartesian coordinates (see figure 2.16). In describing the motion of the atmosphere on our (nearly spherical) Earth, it would be natural to use spherical polar coordinates; however, such a coordinate system and the terms that arise from the curvature of the Earth's surface add unnecessary detail if our primary goal is to describe the motion of a parcel of air over relatively short distances from one moment to the next. For this reason, we use local Cartesian coordinates throughout this book.

Use of the chain rule of differentiation for any quantity f evaluated at the position of the air parcel $\mathbf{x}_p(t) = (x_p(t), y_p(t), z_p(t))$ shows that

$$\begin{aligned} Df(x_p(t))/Dt &= \partial f/\partial t + (\partial f/\partial x)\, dx_p(t)/dt + (\partial f/\partial y)\, dy_p(t)/dt \\ &\quad + (\partial f/\partial z) dz_p(t)/dt \\ &= \partial f/\partial t + (\partial f/\partial x)u + (\partial f/\partial y)v + (\partial f/\partial z)w \end{aligned}$$

since $(u,v,w) = (dx_p(t)/dt, dy_p(t)/dt, dz_p(t)/dt)$ is the rate of change of the air parcel position with time, which is the definition of the wind speed and direction, that is, the velocity $\mathbf{v} = (u,v,w)$.

We can now rewrite all the equations in chapter 2 so that instead we have equations that hold at each grid point in latitude, longitude, and altitude above mean sea level. Richardson developed his practical calculation method based on this formulation of the basic equations, where his particular choices were six hours for the time step Δt and grid spacing Δx, Δy of 200 km and depths Δz that corresponded (approximately) to equal differences between pressure surfaces.

As an example, we write the mass conservation equation from tech box 2.1 as

$$\begin{aligned} D\rho/Dt &= \partial\rho/\partial t + u\,(\partial\rho/\partial x) + v\,(\partial\rho/\partial y) + w\,(\partial\rho/\partial z) \\ &= -\,\rho(\partial u/\partial x + \partial v/\partial y + \partial w/\partial z). \end{aligned}$$

The term to the right of the final equals sign in this last equation is the product of density, ρ, with the total convergence of the wind.

six hours, which replaced the infinitesimal interval in differential calculus. By a systematic application of his finite differences, Richardson broke the undo-able, difficult problem down into a very large number of simpler problems that he could actually solve with elementary arithmetic—the ultimate reductionist success.

The pattern of pixels laid out on the landscape has been described as a checkerboard for obvious reasons. What might not be immediately obvious is that the labeling of the pixels with the letters "P" and "M" reflects nature's game of chess involving the pressure gradient force and its effect on momentum according to the rules laid down in Euler's equations. In the P pixels, Richardson recorded the barometric pressure, moisture, and temperature. In the M pixels, he calculated the (horizontal) momentum of the atmosphere (the product of the wind speed in a given direction and the mass of the air parcel being transported by that wind).

For Richardson to calculate the force that causes the speed and direction of the wind to change in an M pixel according to Newton's laws, he first had to compute the rate at which pressure is changing with location. Thus, increasing pressure to the west will tend to move air eastward, and he computed this pressure gradient from the values held in adjacent pixels. This was done using only simple arithmetic by taking the differences between the values at neighboring pixels, as shown in figure 3.3.

The equations Richardson used, based on Newton's laws of motion, tell us that the change in the wind from one moment to the next is determined by the change in the pressure from one location to another. The change in the eastward component of the wind depends on the difference between the pressures in the P pixels to the east and west; in the same way the northward component of the wind is calculated from the difference in pressure values to the north and south. Consequently, the value of the wind speed in the eastward direction at any one location depends on the values of the pressure two hundred kilometers away on either side.

This scheme works fine until we get to the boundary of the region. Computing changes when the point A in figure 3.3 is at the boundary presents a complication because at the edges of the grid there won't be any adjacent pixels. The ideal solution to this problem would be to cover the entire globe so that there are no edges in the horizontal. But, as we have already explained, this would have led to an impossibly time-consuming calculation in those days without supercomputers. So we need to apply conditions on the variables at the boundaries—including

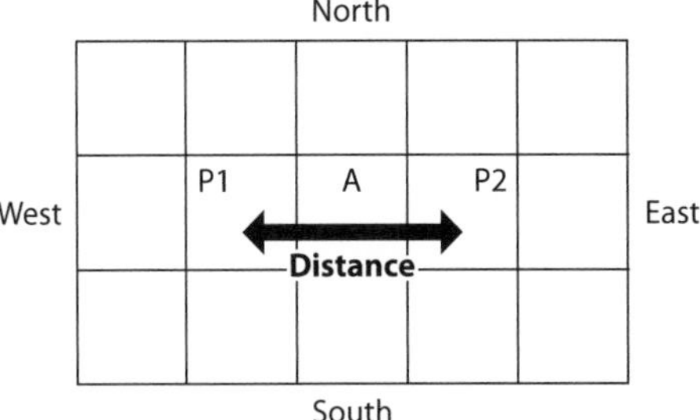

Figure 3.3. A simple example of a computational grid: the values of pressure are stored at the center of each alternate pixel. To calculate the pressure gradient (that is, the rate of change of pressure with location), we simply subtract the value P1 from the value P2 and divide by the distance between them. This would give us the pressure gradient (in the eastward direction) in pixel A. The same principle applies to working out how variables change over an interval of time, by replacing "distance" with "elapsed time."

the pixels at the top and bottom. The lower boundary will be the Earth's surface (land or sea), but an upper boundary also has to be specified because our model cannot extend indefinitely into space.

Richardson skirted the issue about horizontal boundaries by making his forecast for only two locations in the middle of the checkerboard. Thus he made use of the initial values in the boundary pixels but did not try to calculate new values there. This further approximation was entirely reasonable because Richardson was only interested in calculating a six-hour forecast at the center of the checkerboard many hundreds of kilometers from the boundaries. The errors he would introduce by treating the boundaries in this way did not influence the accuracy of the forecast in the center over this limited period of time.

The Anatomy of an Error

Richardson returned from the war in 1919 to work once more for the Meteorological Office. He then wrote a detailed book describing his work titled *Weather Prediction by Numerical Process*, which was published in 1922. In the central M pixel, which covered the area surrounding Nuremburg and Weimar, the surface wind freshened somewhat over the six-hour period. However, in the P pixel over Munich, the pressure rose by 145 mb to 1,108 mb. If this surface pressure had been correct, it would have been a world record! The actual barometric reading on that day was almost no change—or, as weather forecasters say, nearly steady.

It is ironic that this groundbreaking forecast was one of the worst ever, but this failure should be put into perspective. First, the complexities of weather prediction by numerical procedures are such that Richardson estimated he would need sixty-four thousand people to compute a weather forecast for the whole world using an extended version of the model he employed for central Europe. Even then, this army of people would barely keep pace with the changes in the weather; in other words, they would not actually be performing a forecast. In order to do that, we would need more than one million people working around the clock. This is what supercomputers now do in a matter of seconds.

Second, we need to take a closer look at what went wrong. The problem was more subtle, and it was identified and described by Richardson in his book. At the end of his table of results, he wrote: "This glaring error is examined in detail below . . . and is traced to errors in the representation of the initial winds." That is, the problem lay with the input data. A close look at the data confirms this diagnosis and shows us one of the reasons why weather prediction is such a hard problem. The main trouble spot is the part of the calculation where pressure is determined from the convergence of the winds, as we now discuss.

Along each horizontal axis, the wind convergence is calculated as the small difference between two larger numbers. Under these circumstances, even slight errors in the initial wind data can cause large variations in the computed convergence. For example, as shown in figure 3.4b, suppose we have a basic northeastward wind that is slowing down. It is 10.1 kph to the east on the left side of our P pixel, and 9.9 kph to the east on the right side of the pixel. This deceleration gives rise to a horizontal convergence that is equal to $2(10.1-9.9) = 0.4$, where the pixel has sides of unit length. If the measurement of the eastward wind is in error by just 1 percent, say 10.1 is measured as 10, then this convergence will be $0.2 = 2(10-9.9)$ instead of 0.4; the horizontal convergence changes by 50 percent. This disastrous error will be reflected in the predicted barometric pressure; in other words, a tiny relative error in one of the wind speeds becomes a major error in the force represented by the pressure gradient.

So the basic techniques were sound, apart from the sensitivity of the pressure calculation to small errors in the wind and temperature fields, and this sensitivity still requires careful handling in simulations today.

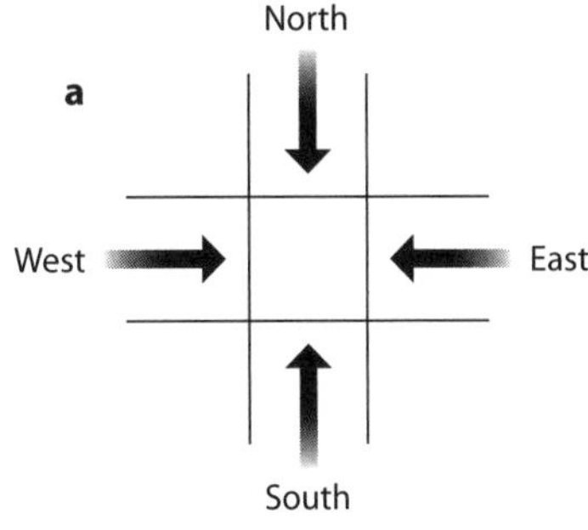

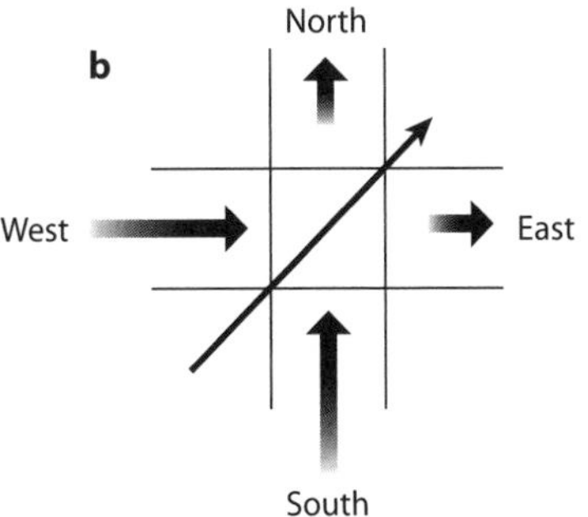

Figure 3.4. In (a) we show a horizontal cross section through a pixel at ground level with the arrows showing incoming winds from the north, east, south, and west. Such winds accumulate air in the pixel, and horizontal convergence measures this accumulation. Since air cannot appear or disappear, either the density increases in the pixel, or the excess air escapes by moving upward. We write –*div***v** for the total convergence in a pixel. This total convergence, which includes the vertical convergence, is usually very small, even when strong winds are blowing. In (b) the longer solid arrow shows a decelerating wind heading in the northeastward direction. First we subtract the central (average) value of this wind from the east–west and north–south components. Then we get the convergence as we did in (a). So a changing wind field over our grid has associated with it a value of the convergence on each pixel.

Indeed, nearly eighty years on, Peter Lynch of Met Éireann in Dublin analyzed Richardson's forecast and confirmed that the error arose from the observed initial state and not from the calculation of the changes that ensued. Lynch used a modern computer simulation to replicate Richardson's forecast (not wanting to endure the months of laborious arithmetic by hand), and he found that by controlling a suspected error in an observation of the air pressure over Strasbourg, Richardson's algorithm does indeed produce a good forecast. The forecasts we now see on the news and weather channels are based on techniques remarkably similar to those that Richardson devised—including such occasional errors.

Although Richardson had lost a battle, he had identified what would turn out to be a successful strategy for winning the war between weather and weather forecasters. The error did not result from a "bug" or error in his calculations, and his mathematical methods have always been recognized as nothing less than a triumph. However, Richardson was a realist and appreciated that his method was of little practical value in the 1920s because of the time it had taken him to compute his forecast.

In the nineteenth century teams of people who were quick and accurate at arithmetic were used to carry out various scientific procedures that could be represented mathematically. In the 1850s the Astronomer Royal employed eight men to calculate the times of the tides, planetary appearances, and other celestial events. Richardson's estimate that a workforce of sixty-four thousand people would be required to calculate the changing weather patterns at a rate that would only just keep pace with the changes was optimistic. Although the scheme itself was quite impractical in the pre–computer era, Richardson was hopeful: "Perhaps some day in the dim future it will be possible to advance the computations faster than the weather advances and at a cost less than the saving to mankind due to the information gained. But that is a dream."

His dream was finally realized when the first electronic computers became available for research use after the Second World War. An image of Richardson's "forecast factory" is shown in the color section, figure CI.4.

Richardson was elected a Fellow of the Royal Society of London in 1926. The Meteorological Office became part of the military establishment after the First World War, so, adhering to his pacifist principles, Richardson resigned. In fact, he did little further work on meteorological problems and, after his death in 1953, his wife Dorothy recalled "a time of heartbreak when those most interested in his upper air researches proved to be the poison gas experts. Lewis stopped his researches, destroying such as had not been published. What this cost him none will ever know!" (Cox 2002, 162).

The legacy of Richardson's work is enormous, and his memory is honored by methods such as the Richardson extrapolation, and by the measure of the critical stability of stratified fluids, such as the oceans and atmosphere, in terms of the Richardson number. Richardson's numerical weather prediction showed everyone that a direct attack on the problem was unlikely to meet with success unless guided by a deeper understanding of meteorology: where would that insight come from? To begin to answer this question, we return to the saga of Bjerknes's trials and tribulations.

The Cradle of Modern Meteorology

While Richardson was beleaguered on the Western Front, life was not much better for Bjerknes, despite the success of his research institute in Leipzig that he had created in January 1913. His difficulties came to a head in the winter of 1916–17 when food shortages and other restrictions in Leipzig forced the Bjerknes family to consider a return to Norway. Coincidentally, moves were afoot in Bergen to create a new geophysical research institute; on March 17, 1917, the organizing committee for this new institute sent Bjerknes an invitation to a specially created professorship to lead a weather forecasting team in Bergen.

During the decade before he returned to Bergen, Bjerknes's career in the universities of Christiana and Leipzig had been the pursuit of his 1904 manifesto to predict the weather using the laws of physics. Bjerknes hired assistants using the Carnegie grant and wrote the first two books of a four-volume project to define the science of meteorology. He spent much of his time evaluating the basic building blocks that such a program entailed, but especially in connecting observations of pressure at ground level to movement of the air currents in the lower to middle atmosphere. The predictions for the "rivers of air" were of vital interest to the developing aircraft industry. Instead of tackling the forecasting problem using numerical techniques alone, as Richardson had gone on to do, Bjerknes believed that a combination of graphical and numerical methods was the only practical way forward.

Bjerknes developed a procedure whereby the state of the atmosphere was represented by a collection of charts on which he plotted the distribution of the basic variables. Each chart displayed the variables at a different level in the atmosphere. The graphical–numerical methods were then applied to construct a new set of charts that described the state of the atmosphere, say, six hours later. In essence, the graph techniques amounted to slick ways of calculating vorticity, convergence, and vertical motion. As with Richardson's scheme, this process could be repeated until a forecast for the next day or so had been made. To draw the initial charts, Bjerknes required observations and international cooperation to gather the data. Campaigns involving balloon flights were organized, and, as we mentioned, one such event provided Richardson with the initial conditions for his forecast.

But Bjerknes's scientific program was about to make a dramatic change of direction. The war had led to severe food shortages in Norway, and in 1916 less than half the amount of grain was produced than was consumed during each of the first two years of the war. Although grain could be shipped in from America, escalating shipping costs (and the consequences of war at sea) led to dramatic price rises.

Domestic production simply had to increase, and the government was forced to intervene in nearly every aspect of food production and supply. In February 1918 a Norwegian national newspaper carried a story about a Swedish plan to make short-period weather forecasts available to farmers by telephone. Although nobody knew how reliable such a scheme would be, it was acknowledged that weather forecasts, even for only short periods of up to a day in advance, could help farmers with planning when to harvest. The article noted that, despite Norway's food crisis, no such plans were even being considered by the government. What was perhaps more alarming was the report that the head of the Norwegian Meteorological Service had played down the possibility of establishing a similar scheme in Norway.

Bjerknes read the article and responded vigorously. He wrote to the head of the Meteorological Service and urged him to reconsider his position, pointing out that he was obliged to consider every avenue to help alleviate the country's food crisis. And, perhaps rather cunningly, Bjerknes pointed out that such a project might be a means to secure much-needed funding for meteorological research. He realized that the war, having created crises that called for a new weather service, might also create the resources needed to study and overcome some of the obstacles to successful forecasting.

Bjerknes's primary goal for the summer of 1918 was to use weather forecasting to increase agricultural production. At the same time, he sought to experiment with possible forecasting practices that could meet the coming needs of aviation. The challenge was to achieve precision and reliability on a scale hitherto never attempted, and conditions for issuing precision forecasts could hardly have been worse.

Before the war, forecasting in Norway relied upon weather data sent by telegram from Britain, Iceland, and the Faroe Islands because this was where most of the weather originated. Now, in a time of war, with German zeppelin raids on Britain and with weather forecasting used in most military operations, this data became secret. Without data from the North Sea and beyond to locate the approaching low-pressure systems, traditional methods of forecasting based on delineating the field of surface pressure could not be used. Bjerknes realized that it would be necessary to create new observing stations around the Norwegian coast to compensate for the lack of data from abroad. He traveled extensively to talk to lighthouse keepers, fishermen, and other keen observers of the weather off the Norwegian coast. Bjerknes not only gained their sympathy and cooperation, he also learned about "real" weather and its "signs." New weather stations were created as part of Norway's U-boat observation network. To ensure that Norway's neutral waters remained free from U-boat activity, a small number of outlook posts, manned by experienced mariners who were well trained in observing weather, were established around the coastline—even in the most remote areas. By the end of June 1918 Bjerknes had established sixty new weather stations, and Norwegians were on the lookout for cyclones as well as submarines—both were affecting food production, and Norwegians were hungry.

But Bjerknes created much more than just new sources of data; he created what was to become a world-renowned school of modern meteorology. The Bergen School, as it became known, resembled a school in more ways than one. The scientists, armed with their rulers, compasses, protractors, and slide rules, sat at desks amid piles of charts and graphs, working according to a strict routine. The address, Allégaten 33, in Bergen, was home to both the school and the Bjerknes family. It was a substantial residential property situated on the edge of Bergen's finest public park on a rise of land near the town center. The house provided splendid accommodation for the Bjerknes family on the ground floor and offices for the forecasting venture on the top floor. A contemporary photo of the school is shown in figure 3.5.

Bjerknes's constant presence and long-established leadership in the field ensured continued stimulus and inspiration for his young colleagues. Out of their labor-intensive hard work emerged a basis for modern weather forecasting.

The Bergen School at that time had four leading figures: Vilhelm Bjerknes, the senior member and the key personality in every way; Jack Bjerknes, Vilhelm's son and only twenty years old when the School was established; and Halvor Solberg and Tor Bergeron, also in their twenties. Although the most striking innovations of the Bergen School, both in working methods and theory, were mainly due to the three younger personalities, these were almost certainly only possible within the context of Vilhelm Bjerknes's well-established research strategy and the inspiring leadership he always seemed to provide.

In the first decade of the twentieth century Bjerknes had mentored three gifted graduates—Vagn Ekman, Bjørn Helland-Hansen, and Johan Sandström—to interpret and apply his circulation theorem to various

Figure 3.5. The Bergen weather forecasters at work (in the Bjerknes family home) on November 14, 1919. On the far left in the foreground we see Tor Bergeron, seated at the desk, with a young student, Carl-Gustaf Rossby, to his left. The gentleman standing with his back to the camera is Jack Bjerknes. To Rossby's left is Svein Rosseland, and seated at the back of the room facing the camera are Sverre Gåsland and Johan Larsen, both clerical assistants. The lady seated at the desk to the right is Gunvor Faerstad, who was responsible for receiving meteorological observations by telephone and entering the data on the charts. © ECMWF. Reprinted with permission.

challenging examples of ocean flow. Together this group produced the cornerstones of theoretical oceanography. It is remarkable that Bjerknes had similar energy and talent a decade later for these three new students who were destined to lay the foundations of modern meteorology, as we describe in the following section. And that was not all; a graduate by the name of Carl-Gustaf Rossby would also spend nearly two years at Bergen en route to the United States. Rossby would play a decisive role in transforming American meteorology in the 1930s and 1940s. Rossby would go on to fully exploit the power of Bjerknes's circulation theorem in producing the first quantitative explanation of large-scale weather patterns (as we discuss in chapter 5).

The graphical methods developed by Bjerknes and his new protégés encapsulated an important "alternative view" of the equations we presented at the end of chapter 2. From the outset, Bjerknes was driven in his quest by the remarkable applicability of his circulation theorem. This theorem, which we described in tech box 1.1, is a consequence of the fundamental governing equations. Richardson developed techniques to solve these equations in their finite-difference form: implicit in Bjerknes's graphical techniques were methods that made use of convergence and its relation to changes in the circulation. Bjerknes had realized he did not have the means to directly calculate convergence accurately enough, but he could calculate the circulation. We now turn to explaining how these ideas and concepts emerged.

From the Sea Breeze to the Polar Front

Bjerknes's ideas about atmospheric circulation were first calculated by using paths that made loops lying within a vertical cross section of the atmosphere, and were explained in his 1898 paper (see tech box 1.1 on the sea breeze). The next major breakthrough occurred when Bjerknes's Bergen team considered instead these paths lying on nearly horizontal constant-density surfaces. Now flows such as those depicted in figure 3.8 and 3.12 could be studied—images of weather that are familiar to us today.

So how do we make the transition from theory to practice? Bjerknes was concerned with circulation around large paths that encircled the region of interest. Because we cannot stir the atmosphere and observe the consequences of different methods of stirring, laboratory experiments are constructed to measure and analyze the behavior of rotating fluids. Theoreticians also think of "toy experiments" such as stirring a cup of coffee or watching water flowing out of a drain hole, as shown in figure 3.6.

How can we form a mental picture of what an appropriate path is and what circulation means? Consider such a flowing liquid escaping down the drain hole, and ask the question: is there a net circulation of fluid? Here we quantify the similarities between the flow in a smoothly stirred cup of coffee and that of the water draining out of the bath. First, position a circular loop near the lip of the cup or around the drain hole of the bath. Suppose that we ignore the liquid and its motion everywhere except inside a hypothetical tube of uniform bore, which is placed around the circular loop of radius R, as shown in figure 3.7.

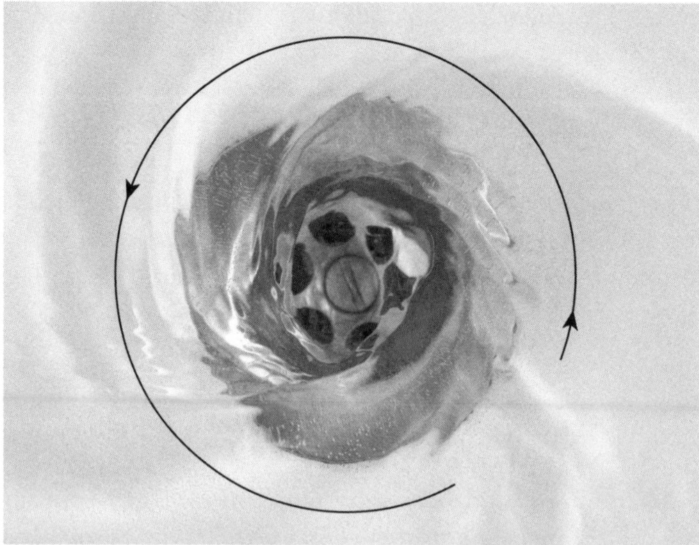

Figure 3.6. Swirling water flows out of a sink or bath by means of a central drain hole. As fluid parcels rotate and approach the drain hole, they speed up to conserve the circulation. © Pavel Losevsky / 123RF.COM.

What will happen to the fluid in the smooth tube now that we have disconnected it from its surroundings? The fluid inside the tube should keep on moving steadily because of the net momentum of each fluid parcel moving around the circular loop. We define the circulation as the resulting average speed of the liquid in the tube multiplied by the distance ($2\pi R$) each fluid parcel travels around the tube.

The experiment described here involves thinking about fluid motion in a more macroscopic context, that is, "in the large." But the equations governing fluid motion—the same equations Bjerknes and his team were grappling to extract information from—are expressed in terms of tiny fluid parcels. So to connect these concepts, we need to apply the ideas of circulation to each fluid parcel. To this end, we contract the large path into a small circular path around a typical fluid parcel. This allows us to calculate another fundamental measure of the swirl of the fluid, the point vorticity, defined as the amount of circulation around a small loop in the fluid divided by the area enclosed by that loop, as described in the caption to figure 3.7.

If a disc-like mass of fluid, such as that shown in figure 3.7, is swirling around at a constant rate so that the angular velocity of all fluid parcels

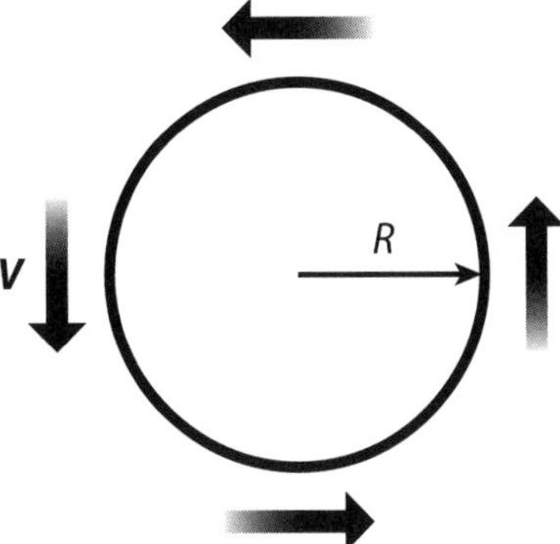

Figure 3.7. Consider a circular path (or loop) centered on a point vortex. The circulation, C, is equal to $2\pi RV$, where R is the radius of the loop and V is the speed of the flow (shown by thick arrows) around the loop, which has length $2\pi R$. The vorticity is the circulation around the loop divided by the area (πR^2) enclosed by the loop. Then vorticity is $C/(\pi R^2) = 2\pi RV/(\pi R^2) = 2V/R = 2\omega$, where ω is the angular velocity of the vortex flow around the loop. If we were to drop a piece of cork into the vortex, then the cork would travel on the loop around the vortex with an angular velocity equal to half the vorticity. The sketch represents an idealized disc of rotating atmosphere, such as in figure 3.8, or in figures 1.3 and 1.10.

is the same, then the fluid traveling around at the edge of the rotating disc has more circulation than the fluid moving in smaller concentric loops. In this way circulation around a loop measures the total strength of vorticity inside the loop. Larger loops have more swirling fluid in them and add up more "point vorticity." In practice, the speed of a bath plug vortex is usually faster near the center so that the circulation of various loops (around the plug hole) is nearly constant.

Over the past 150 years, no practical way has been found to directly measure vorticity or circulation—unlike pressure or temperature, we cannot buy a "circulation meter" or measure the vorticity directly from a satellite—but we can calculate circulation from the velocity of the flow. As Helmholtz had noted, vorticity is unaltered in perfect fluid flow unless something significant happens to change it. And convergence is just such a thing. In fact, local convergence of the horizontal wind is one of the few things that changes local vorticity because it is usually connected with movement of air in the vertical. By adding all the local convergences, we get the related Bjerknes circulation result, that net convergence inside a larger loop in the fluid is associated with a change in the total circulation around that loop, as tech box 3.2 explains in more detail.

Figure 3.8. Hurricane Dora was photographed from NASA's Terra satellite on July 20, 2011, when it was over the coast of southwestern Mexico. The swirling spirals of warm moist air are shown by cloud. In fact, without knowing the direction of the air movement, we might erroneously see this as a "bath plug vortex," with the air swirling in and draining or disappearing down the calm "eye," which is clearly visible in the center of the picture. Air actually rises near the eyewall, and swirls and spills out at higher altitudes. This picture shows dramatic motion of the atmosphere on a large scale of several hundreds of kilometers. Reprinted courtesy of NASA.

With new data sources and new ideas, Bjerknes and his team were well placed to undertake detailed analyses of weather systems over Norway, and they were motivated by the need to help farmers and fishermen. Following in his father's footsteps, Jack Bjerknes became a key member of the small research group. (Later in the 1940s he would help create a university department of meteorology in Los Angeles, where our key protagonist of chapter 6 learned his trade.)

Tech Box 3.2. The Backbone of Weather I: The Vorticity Equation

Weather systems are often observed to change systematically. As we have just discussed, vorticity is directly related to circulation, and in tech box 1.1 we explained how changes in circulation are related to systematic changes in the winds, such as the sea breeze. We now start from the governing equations of tech box 2.4 and manipulate them to focus on vorticity. We make one further assumption: that the density is well approximated by its hydrostatic averaged value $\rho(z)$, as described in tech box 2.2.

So we start with the horizontal wind equations with ρ replaced by its averaged value $\rho(z)$:

$$\partial u/\partial t + u\,(\partial u/\partial x) + v\,(\partial u/\partial y) = f_c\, v - (1/\rho(z))\partial p/\partial x,$$
$$\partial v/\partial t + u\,(\partial v/\partial x) + v\,(\partial v/\partial y) = -f_c\, u - (1/\rho(z))\partial p/\partial y.$$

Here we ignore the vertical motion, w, and for the present purposes we ignore any significant variations in temperature. The terms on the left-hand side are the total derivatives $D_H u/Dt$ and $D_H v/Dt$, expressed in terms of ordinary partial derivatives as explained in the previous tech box and where the subscript "H" denotes the fact we are considering motion in the horizontal.

We focus on the vorticity $\zeta = \partial v/\partial x - \partial u/\partial y$ by taking the $\partial/\partial y$ derivative of the first equation and subtracting it from the $\partial/\partial x$ derivative of the second equation. This has the side effect of eliminating the pressure gradient term, as first noticed for fluids with constant density by Helmholtz.

By collecting terms and using the fact that f_c varies only with y (latitude), we take the result of the above calculations

$$\partial\zeta/\partial t + u\,(\partial\zeta/\partial x) + v\,(\partial\zeta/\partial y) = -\,(f_c + \zeta)(\partial u/\partial x + \partial v/\partial y) - v\,(\partial f_c/\partial y),$$

and rewrite it as follows:

Proportional change of total vorticity $(f_c + \zeta)$
in the horizontal direction =
$(1/(f_c + \zeta))D_H(f_c + \zeta)/Dt = -\,div\,\mathbf{v}_H$ = horizontal convergence.

Practical weather forecasters estimated the horizontal convergence with increasing skill throughout the twentieth century. Because changes to f_c were known, forecasters could then estimate changes to ζ. Thus, they learned to link changes in the convergence to changing weather as the cyclone developed and migrated, mostly in an eastward direction.

Earlier in 1917 in Leipzig, Jack Bjerknes had begun a study of so-called lines of convergence. By transforming the equations and applying the theory of vortex motion to these patterns, he arrived at an equation that related the movement of the lines of convergence to the horizontal rate of change of the vorticity field, as we outlined in tech box 3.2. When air converges at low levels, it is accompanied by vertical motion, which causes upper-level divergence, and surface pressure falls. Jack Bjerknes deduced that tracking the lines of convergence of the wind tells us about changes in the pressure and, hence, the attendant weather. Up to this point, understanding and predicting pressure patterns had been a cornerstone of forecasting, and this is still what is shown on a weather map to this day. Vorticity and convergence might now displace pressure from its primary role, as explained at the end of tech box 3.2.

After the move from Leipzig, Jack Bjerknes's work at Bergen again focused on the lines of convergence. But now, with much more weather data available from the observers stationed around the Norwegian coast, he identified patterns in rainfall associated with patterns in the airflow.

In October 1918 Jack Bjerknes wrote a paper that was to revolutionize meteorology. The paper was entitled "On the Structure of Moving Cyclones," and it attempted to explain the patterns observed in the previous summer's weather maps. The findings were to lead to the conceptual models of warm and cold fronts—he called these "steering lines" and "squall lines" as shown in figures 3.9 and 3.10—and they became linchpins of modern meteorology. The name "front" was adopted later, which originated with the notion of a battlefront between the competing effects of warm and cold air—an idea that would have been familiar to all at that time from the maps of fronts in the First World War that appeared regularly in the newspapers.

Jack Bjerknes had overturned the previous century's view of a cyclone as a smooth, nearly circular object and transformed it to a towering structure of differing warm and cold parts swirling through the atmosphere. The new model seemed to possess an asymmetrical thermal structure consisting of a distinct "tongue" of warm air bounded by colder air, as shown in figures 3.9 and 3.10. By identifying the steering lines, he hoped to be able to predict where the cyclone would go next; by finding the squall lines, he could predict where the rain showers would occur.

Plotting the flow lines, or streamlines, helped with the problem of forecasting for Norway. If lines of convergence of the streamlines could be identified as they approached the observation posts on the coast, then the movement and speed of the convergence could be calculated, making it possible to issue reliable forecasts of rain to farmers in advance of the wet weather. Jack Bjerknes's diary for the summer of 1918 reveals his increasing familiarity with the relationship between lines of convergence and low-pressure systems. His daily routine was based

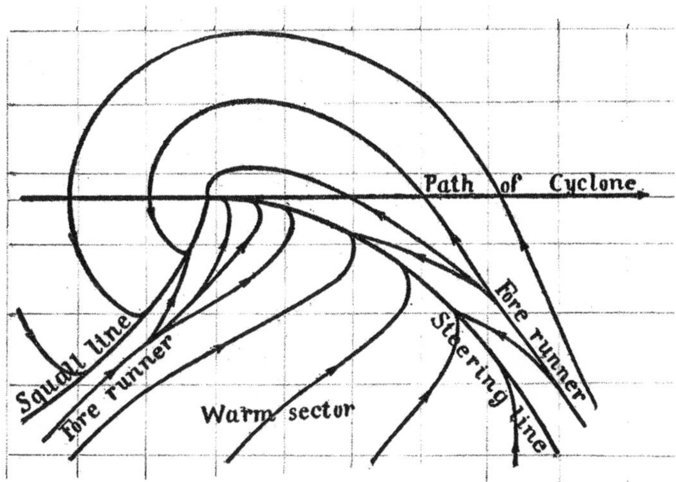

Figure 3.9. A figure taken from Jack Bjerknes's paper, "On the Structure of Moving Cyclones," published in 1919. Early in the paper he writes, "Both the steering line and the squall line move according to the law of propagation for lines of convergence." The lines of flow, or streamlines, are shown, and the convergence of these lines could now be related to vertical motion. In turn, this facilitated an understanding of where rainfall could be expected in such a weather system. Nearly a century after the Espy–Redfield controversy, cyclones acquired an internal structure. *Monthly Weather Review* 47 (1919): 95–99. © American Meteorological Society. Reprinted with permission.

around receiving data from the observation stations at 8:00 a.m., which had been sent by telegraph from around the country to Bergen. He then worked with his colleagues in plotting this data and drawing up charts. By 9:30 a.m. Jack was issuing forecasts for the period up to twelve hours ahead to various regional centers. Local farmers could then pick up the forecast information. An additional set of observations received in the early afternoon served as a check on the predictions that had been issued for that day.

This daily routine of sifting through a mountain of observations and plotting weather patterns was the key to the Bergen group unlocking the secrets of low-pressure systems. Figure 3.10 is an enlargement of Jack Bjerknes's description of the typical weather in a vertical cross section of the atmosphere. An actual example from modern weather forecasting is shown in figures 3.11 and 3.12.

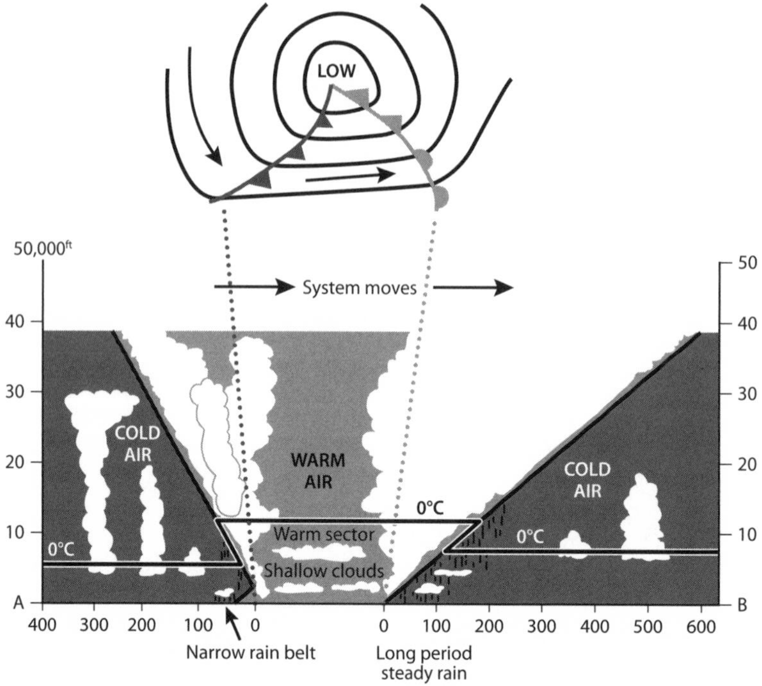

Figure 3.10. A Bergen School idealization of a low-pressure system. Jack Bjerknes realized that rain was usually located near these abrupt changes from cold to warmer air, shown as sloping lines in the lower slice through the atmosphere. Now known as fronts, there is usually more persistent rain on the shallower warm front and sharper outbursts on the steeper cold front. © ECMWF. Reprinted with permission.

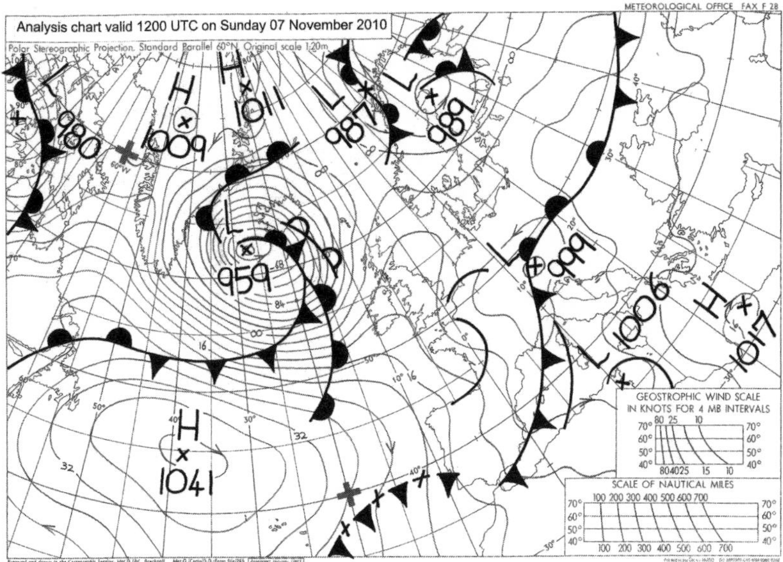

Figure 3.11. A forecaster's chart showing a significant low-pressure system heading in an approximately eastward direction toward Iceland. Such lows are developing through their life cycles, and creating local weather that can be predicted using Jack Bjerknes's ideas. The warm and cold fronts of the low-pressure system (with the lowest pressure of 959mb indicated at its center) are shown by the now widely used, semicircular discs (warm front) and triangles (cold front): note the slight change in the direction of the isobars (the solid lines) at the warm front. The Bergen School developed the use of these synoptic charts (and the plotting of various fronts associated with low- and high-pressure centers) in weather forecasting. (This chart was redrawn by hand by Steve Jebson.) © Crown Copyright, Met Office. Reprinted with permission.

Tor Bergeron's first major contribution came in November 1919. Several times during his forecasting work, Bergeron detected features on his maps that seemed anomalous according to the concept of lows as expressed by Jack Bjerknes. Whereas Jack Bjerknes's model allowed only the possibility of separation between what were to become known as the warm and cold fronts in the southern part of the tongue of warm air, Bergeron suspected that the cold front could catch up with the warm front. They might even come together in some distinctive way. If we look at figure 3.10, we might imagine the cold air to the left catching up with the cold air on the right, thus lifting the warm air from the ground. Analyzing local Norwegian weather data for November 18, 1919, Bergeron expressed his idea in a confident drawing where these two lines actually

Figure 3.12. This satellite image taken over the North Atlantic Ocean shows the cloud patterns on the day of the forecast of figure 3.11. The blanket of cloud over the western Atlantic, Ireland and Scotland, is associated with warm moist air, and the warm front. The thinner, stippled, cloud pattern sweeping down from the north between Greenland and Iceland is the colder, drier air, and the cold front. © NEODAAS / University of Dundee

did merge, and this opened up the idea of an "occlusion," which occurs when a warmer air mass becomes separated from the ground and enclosed by colder air, as shown in the evolving sequence (parts d and e) depicted in figure 3.13.

The way now became clear for seeing that lows or cyclones do not have one static structure; they have an evolving structure corresponding to a life cycle of birth, development, and decay, which typically lasts for one to two weeks. So striking were these advances that Vilhelm Bjerknes began preparing a paper entitled "On the Origins of Rain"—a title inspired, of course, by Charles Darwin's famous paper on the origin of species—but the paper was never completed. Real weather seemed to vary just too much! However, the concept of these cyclones, continually traveling through the atmosphere as they marked out their life cycle, shedding rain at key maturation stages, was to transform meteorological observation, thinking, and prediction.

The third major constituent of the research of the Bergen School was the discovery made by Solberg during February and March 1920, when Jack Bjerknes's model was attracting worldwide attention. Solberg

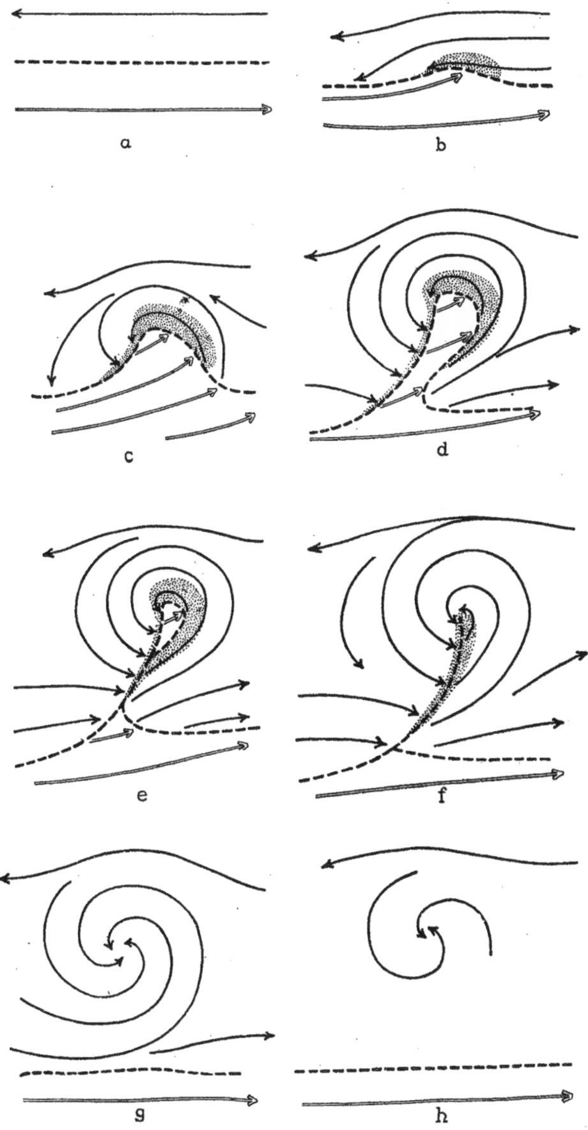

Figure 3.13. The plan view of the life cycle of a cyclone is shown, based on an idealized polar front, where lines of latitude run eastward from left to right. The solid lines show the flow of marked air parcels, whereas the dotted lines indicate the fronts that meteorologists mark on weather maps. This battleground between the warmer midlatitude air and the colder air of the polar region, as the birthplace of midlatitude weather systems, was a transforming idea for meteorologists. The polar front is shown in (a), and the beginning of a low-pressure system is shown in (b). Then (c), (d), and (e) show the cold part of the front catching up on the warm part of the front as it matures. Finally, (f), (g), and (h) show the mature cyclone gradually decaying back to the original state. *Monthly Weather Review* 50 (1922): 468–73 . © American Meteorological Society. Reprinted with permission.

consolidated and completed the emerging ideas of a cyclone and its life cycle by identifying a global front stretching all the way around the polar circle, which delineated warm air to the south from cold air to the north. Known as the "polar front," this appeared to be the favored location for such lows and gave rise to the notion of families of cyclones circling around the globe, like beads on a string of latitude. Vilhelm Bjerknes immediately recognized the value to practical forecasting—if one could identify the point in a cyclone's life cycle that had been reached by a weather system crossing the North Sea, then such information would give important clues as to what was coming next in the continuing advancement of weather from over the North Atlantic.

The main accomplishment of the Bergen School lay in the introduction of realistic models of the structure of such lows. These models accounted for observed meteorological elements and were directly connected with the weather that the forecasters sought to predict. Vilhelm Bjerknes and his protégés were the first to describe weather in a systematic and scientific way. The key patterns shown in figures 3.10 and 3.13 had been distilled from thousands of observations. The conceptual models of the Bergen School then made useful predictions of the weather of the middle latitudes for a day or two ahead. By the mid-1920s the criticisms leveled at forecasters in the middle of the nineteenth century, that weather forecasting was based only on guesswork using past experience, had been overcome. Once again, better observations and their analysis led to better scientific ideas and, hence, better prediction—Descartes's proposals for a scientific method were yielding, centuries later, lifestyle revolutions for mankind throughout science and, here, transforming meteorology.

A Top-Down View Paves the Way to Progress

Thinking about warmer and cooler air masses meeting at fronts led to the development of *air mass analysis*. By the mid-1920s the Bergen meteorologists had recognized that large bodies of air were marked out by physical characteristics, such as warmth or dryness. This physical identity of the air mass usually arose from its life history. Air masses with an identity form whenever an extensive portion of the atmosphere remains for sufficiently long periods over a broad (often countrywide) region possessing fairly constant sea or ground surface properties. The body of air thereby attains a “thumbprint” (or identity) from its environment, which may be cold and dry over a desert or a tundra in winter, or warm and moist from air masses that sit over the Gulf Stream, for example.

When the air mass finally moves away from its source region and travels over surfaces with different properties, the equilibrium previously established is disrupted; this is when changeable weather can start to appear. For example, when a moist, warm pool of summer air from the Gulf of Mexico moves over the Midwest, which is hotter than the air itself, the hot ground surface heats the warm moist air from below, causing upward air currents that in turn lead to the formation of clouds and heavy rain. At upper levels, each air mass possesses particular persistent properties that make it possible to classify its origin and trace its daily movements and physical changes. Regularly each summer, dust from the Sahara Desert is transported thousands of kilometers by upper air movement until it finally rains out in far-flung places such as the Amazon basin and northwestern Europe.

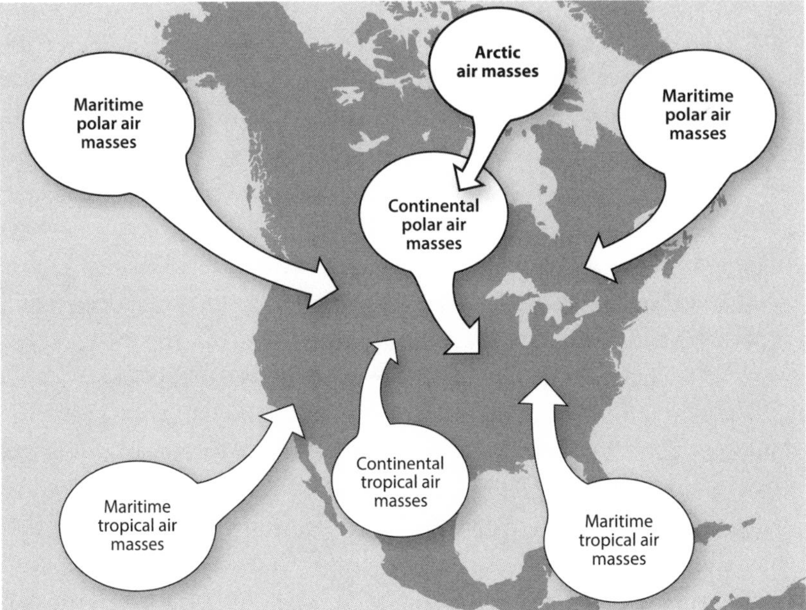

Figure 3.14. This sketch shows the traditional view of the origins and classification of typical air masses that move across the North American continent.

Air mass analysis has value for weather prediction because air masses from known regions possess specific characteristics that, depending on where they travel next, will lead to predictable weather. The great monsoons of southeast Asia, and especially India, are remarkable examples of this. Warm moist air from the Indian Ocean blows across the very hot and drier subcontinent bringing great deluges of rainfall. Another striking example is the cold change that arrives in the northeastern United States from Canada. In spring the warm, moist, Gulf of Mexico air penetrates northward up the midwestern United States against the cold Canadian "shield air," and this is often associated with blizzards and ice storms.

Predicting the characteristics of an air mass by determining both its source or origin as well as its path or trajectory allows forecasters to suggest the weather that is most likely to occur as that air mass rolls on. Weather fronts, typically the boundaries between warmer and colder air, appear routinely on weather maps around the world. However, Bergen School methods were slow in being adopted by most countries' forecasting agencies, with much opposition from the traditional forecasters in the 1920s and '30s. The U.S. weather bureau only started using air mass analysis routinely in 1941.

Worrying about rainfall over the next twelve hours spoiling the harvest in Norway and making food shortages worse had led Vilhelm Bjerknes to realize that weather development all around the planet in this midlatitude zone would ultimately affect the local weather in Bergen. A century later, satellite images allow us to see how interconnected the global weather is. Various major weather bureaus now routinely compute next week's weather for the entire planet.

The Bergen School local forecasts were so useful that fishermen and sailors lobbied the Norwegian government in 1920 to keep the Bjerknes meteorological institute in Bergen. Norway had plunged into a long and major postwar financial depression in 1920. To reduce government expenditure, the Norwegian Ministry of Church and Education in 1924 recommended the closure of either the main weather center in Oslo or that of Bjerknes and his team in Bergen. Naturally, the capital city said their weather center must remain open, especially with the onset of aviation and the planned airship route from London to Oslo and Stockholm.

But along Norway's west coast shipping, harbor authorities, local businesses, and especially fishermen (who said that weather forecasting was the best thing the state had done for them) all argued that the Bergen Institute should remain open. Local newspapers and members of parliament joined in the campaign, and in the end, to Bjerknes's delight, both bureaus were retained. Later Bjerknes would say that of all the forms of scientific recognition that he had received, none gladdened him more than those responses from the fishing community around Bergen.

Some twenty-five years later, by the end of World War II, most of the world's meteorologists were using some form of Bergen School methods, and throughout the latter part of the twentieth century, fronts and cyclone-based weather charts became commonplace in the newspapers and other media outlets. "Rather amazingly," wrote Friedman in 1989, "Norwegian television and Oslo newspapers still avoid fronts on their weather maps." Politics involves power, money, and egos. Many weather services took a lot of converting to the Bergen view as their directors juggled the competing interests of their staff and their countries.

To illustrate this intransigence, we take one early noteworthy example. On October 22, 1921, a storm passed over Denmark and Sweden heading toward Russia, but the Danish weather service had stopped issuing storm warnings on the Danish north coast, and ships and fishing boats were devastated there. When the Danish government learned that the Bergen practices had correctly predicted the storm's intensification, they ordered their weather service to learn and adopt Bergen methods. The head of the Danish meteorological institute did visit Bergen, and did listen to their ideas, but when he returned to Copenhagen he refused to accept or implement the Bergen School approach. It would take time to convert weather services from their empirical methods, rooted in the nineteenth century and use of the telegraph, to use of new scientific methods. For many it took the growth of the airline business in the 1930s and the advent of the Second World War to make this change. The Bergen School was destined to influence meteorology for many decades to come.

By the beginning of the 1920s, Bjerknes and his team were producing astonishing theoretical and practical results, and news of their success spread quickly throughout Europe and America, even though their methods were not adopted as rapidly. Despite the huge advances, Bjerknes was frequently dismayed that success had not been achieved according to his original manifesto. True, he was using the properties of vorticity, circulation, and convergence to infer how weather systems were developing, but the graphical methods he used to extract the information from the equations were inexact when compared to the precise methods of calculus.

It is important to realize that behind the success of Bjerknes's approach lies a key to solving many mathematical problems. Instead of tackling the equations of heat, moisture, and motion directly (as Richardson had tried to do), Bjerknes's team transformed the problem into a set of equations that control weather on the larger scales we are typically interested in when drawing synoptic charts, such as those shown in figure 3.11. That is, instead of computing how the winds and temperature change in detail above each hill and town, they computed how convergence and vorticity change on the larger regional or national scale and then used these predictions to make more general predictions for the average temperature and rainfall in each locality.

The classical physics of the motion, heat, and moisture evolution of air parcels tells us how the basic elements that make up a weather system change. This is our *bottom-up* perspective, as calculated by modern supercomputers using methods very similar to the ones introduced by Richardson and as observed in each locality by an individual weather station. But by studying quantities such as vorticity and convergence in the air above the countryside of a state or region, we gain our *top-down* view: we then infer how the weather pixel variables are controlled as part of an entire weather system as observed today by satellites. Such top-down rules constrain or control the otherwise many different possible independent relationships between the local elements, and they enable us to deduce qualitatively useful facts about patterns in the weather without solving for, or even knowing, all the detail. This type of approach becomes especially useful when we realize there will always be gaps in our knowledge of even the most recent state of the atmosphere, and of many of the detailed physical processes.

Such top-down rules as described in tech box 3.2 are a consequence of the basic laws in tech box 2.3—we have not added any new information in the process of transforming our problem. Such procedures and the principles that emerge from the equations frequently allow us to extract useful information without having to solve differential equations for all the detail either explicitly or precisely. This is a bit like being able to identify a cyclone without having to look at all the individual clouds. Bjerknes and his team made the first pioneering steps toward probing the mysteries of weather-related fluid motion using these techniques.

While the theories propounded by the Bergen scientists were being broadcast with almost evangelical zeal, Richardson's calculation, meticulously documented in his book published in 1922, was a sober reminder of the ultimate difficulty of implementing Bjerknes's program to turn weather forecasting into a problem of exact science. After Richardson's failure while explicitly carrying out the Bjerknes program, the Richardson procedures were effectively ignored for several decades. However, with hindsight, perhaps Richardson's only real error was in assuming that his dream of numerical weather prediction would not be fulfilled until the "dim and distant future." His dream would be fulfilled, and the first forecasts were published in the research literature within thirty years of the publication of his book, but it would all happen in a way that Richardson and Bjerknes could never have imagined.

It has been said that Vilhelm Bjerknes made history, but not the history of his initial choosing. Ironically, Bjerknes ended up with innovative practical solutions to understanding and forecasting weather rather than practicing what he had originally preached: the direct use of the equations of motion and of heat and moisture in the problem of weather forecasting. What Bjerknes had created was an inspired school of gifted scientists committed to his ideas, and those ideas were carried to the next generation of both theoretical and practical weather forecasters. Their discoveries and methods—perhaps we should refer to the latter as improvisations—were not only to become part of every forecasters' basic training for many decades to come, but the structures and patterns they identified were to provide a challenge for the next generation of theoreticians whose job it was to come up with a mathematical theory explaining these phenomena.

The theoreticians who followed were well aware of Bjerknes's original triumph; they recognized that the transformation of the basic equations into laws governing the circulation and motion of vortices held the key to further progress. On the other hand, what was really needed to realize Richardson's dream was as yet undreamed of—the ability of mankind to routinely and quickly carry out mind-bogglingly large amounts of arithmetic. This would need the technological and mathematical revolution of electronic computers, brought forward by certain needs of the Second World War. But the theoreticians' preoccupation with convergence, circulation, and vorticity along with the latent need for powerful computers are symptomatic of a fundamental obstacle to using the equations of heat, moisture, and motion for all they are worth. It is not simply that the problem is very large.

To get to the heart of the matter and understand this obstacle to weather forecasting, we next take a closer look at the mathematics that expresses the laws that govern our weather and climate. What is it that makes weather prediction for the next week so difficult? If we understood the physical processes, if we had complete knowledge of the state of the atmosphere, and if we solved our equations perfectly, could we then forecast forever? It turns out that embedded in the evolution of these air masses as they are transported by the wind is a mathematical process that allows chaos to enter weather.

FOUR

When the Wind Blows the Wind

Each year the winter solstice is predicted and the Sun returns to the same position in the sky. Why does the winter storm not also exactly repeat itself? In terms of gales, wind chill, rain, or snow, winter storms of the middle latitudes are broadly similar but yet ceaselessly varying. The Bergen School's attempts to understand the origins and classify the similarities of cyclones would never answer the question of precisely when and where next winter's major storm would strike. With the enormously greater calculational ability of a modern computer, would Richardson's forecast procedure provide the answer? In this chapter we discover what is in the very rules themselves—in the seven fundamental equations—that makes forecasting so difficult and weather so interesting.

Getting to the Heart of the Matter

In 1928 the English Astronomer Royal, Sir Arthur Eddington, made three predictions for 1999: he predicted two plus two will still equal four; he predicted a total eclipse of the Sun on Wednesday, August 11, which would be visible from Cornwall in southwest England (a very rare event); and he predicted that we will not be able to forecast the weather for the year ahead. Well, we all still agree that two plus two equals four, and there was indeed a total eclipse of the Sun on August 11, 1999, which made headline news in England. Meanwhile weather forecasting for even *five days* ahead remains a challenge for the modern weatherman—even with supercomputers that had not been dreamed of in 1928.

Eddington did not know about chaos, at least not in the way we understand the terminology today, but he did realize that because there are so many factors that influence the weather, then surely we shall never be able to observe and measure all these influences, and this implies an ever-present deficiency in the forecasting process. Predicting the future state of the atmosphere and oceans involves working out many complicated processes with sensitive and subtle feedbacks.

In his grand vision set forth in 1904, Bjerknes contrasted his rational approach to forecasting with the problem of predicting the exact motion of three planetary bodies whose gravitational forces strongly affect each other. He acknowledged that this seemingly innocuous calculation of planetary motion was beyond the capability of contemporary mathematical analysis. The three-body problem had been a hot topic in the 1880s and '90s, and Bjerknes had rubbed shoulders with the world's leading experts in this area, including Poincaré. Although Bjerknes had reduced the problem of describing and predicting the weather to just seven equations or rules, in seven variables that describe each weather pixel, this was much more involved than the stability of our solar system. And the "seven equation count" is somewhat misleading as to what is really involved since each weather pixel interacts with all its neighbors.

Bjerknes foresaw difficulties back in 1904: in section 3 of his paper he states, "Even the computation of the motion of three mass-points, which influence each other according to a law as simple as that of Newton, exceeds the limits of today's mathematical analysis. Naturally there is no hope of understanding the motion of all the points of the atmosphere, which have far more complicated reactions upon one another." As we remarked in chapter 1, using seven numbers to specify seven basic variables at every intersection of lines of latitude and longitude at each degree of meridian means that there will be $360 \times 180 \times 7$ numbers involved, and then we need to do this at a sufficient number of height levels to describe our weather. So Bjerknes's seven equations in seven unknowns eventually become more than ten million equations in ten million unknowns, and each one is potentially as difficult to solve as the "three-body problem." Is this the nub of the difficulty? Are there just too many unknowns and too many equations?

No. We should not attribute the problems of accurate weather forecasting to the sheer scale of the calculations in terms of the numbers involved; that would miss Bjerknes's point, which he made when alluding to the intractability of the "three-body problem." The sheer number of variables certainly isn't the root of this difficulty.

Behind the physics of the atmosphere is a phenomenon that challenges even the most powerful supercomputer simulations we use in forecasting today, and it will continue to challenge us in the future, no matter how powerful supercomputers become. It has been known about for centuries: it is present in Newton's law of gravitation, and Newton himself realized that it will manifest itself to anyone who attempts to apply his law to completely predict the motion of three or more planets in orbit around each other. Bjerknes and Richardson were certainly aware that it creates huge difficulties for numerical weather prediction, but they were unaware that it places one of the most important constraints on our abilities to forecast the weather for much more than a week or so. The important terms in the fundamental equations of chapter 2, those that feed back information about the solution into the very processes that determine the solution, have the property referred to mathematically as *nonlinearity*.

Much interesting behavior in science is attributable to nonlinearity in its various forms. Nonlinear processes in weather underlie the reality, the challenge, and the occasional surprise in terms of a storm that wasn't forecast. Because of the importance of this concept, we first spend a few pages on the essence of nonlinearity before returning to feedback behavior in weather.

Nonlinearity is a mathematical term logically meaning "not linear," so naturally we first describe the simpler situation of linear behavior—usually known as "linearity"—which amounts to a special type of relation between cause and effect. Linearity is the key feature behind basic science and the technological revolution that swept Europe and the United States in the nineteenth century. Linearity is well understood in mathematics, and computer programs are constantly improving in their ability to solve very large linear problems.

So what is this linear world? Suppose a definite cause, such as a pulling force, has a definite effect, such as stretching an elastic string by a certain amount. Then this behavior is linear if increasing the cause by a given fraction always increases the effect by the same given fraction; that is, doubling the force here doubles the stretch. Linear models were discovered to be amenable to explicit mathematical solution, which was

Figure 4.1. Ripples like these on a pond can be described accurately by linear theory. © Vladimir Nesterenko / 123RF.COM.

a major breakthrough. Triumphs of linear models include explanation and prediction in basic acoustics, vision, and radio, and in television and mobile phone signal technology. Wave motion (tides and wind driven) on water surfaces also have a linear theory (see figure 4.1). But if the waves start to become big, and especially if they break, then linear theory no longer holds. Simple linear theory breaks down and nonlinear behavior takes over (see figure 4.2).

To continue with our discussion of the term "linear," we switch attention from the elastic string to a volume of (compressible) gas in a cylinder, as in an idealized bicycle pump. We describe the effect of adding an extra force ΔF to an existing force F on the handle of a bicycle pump with its other end blocked (see figure 4.3; the Greek letter Delta, Δ, is commonly used in calculus to denote a small change of a variable). The force is directly related to the pressure of the gas trapped in the pump. The consequence of the additional force is to compress the gas and cause a corresponding increase in density (because we assume the piston smoothly moves in until the pressure in the gas balances the pressure exerted by the piston). The gas law of chapter 2 is linear, assuming that the temperature is kept constant; this follows because the causes

Figure 4.2. Understanding the surf means grappling with nonlinearity. It is very difficult to simulate the spray zone accurately using a computer model. There are many different ways that linear descriptions fail; we focus on those of relevance to our atmosphere.

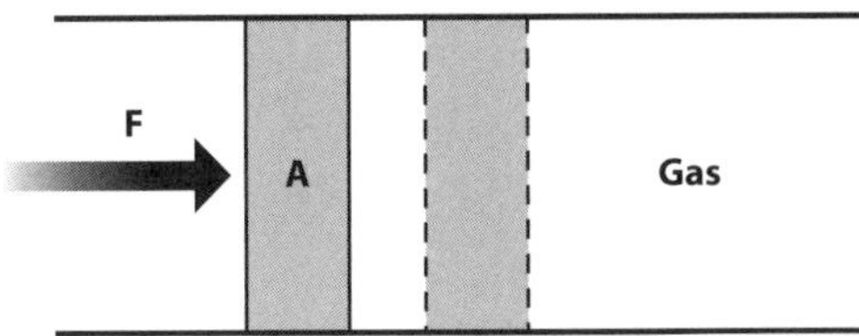

Figure 4.3. In this drawing of a section of a bicycle pump, F represents the compressive force on the gas that is provided through the handle to the piston. A is the surface area of the piston that touches the enclosed air. The dashed lines show where the piston moves to when the force is increased from F to $F + \Delta F$.

(the different pressures) produce the effects (the different densities) in our rule, as we now describe in more mathematical detail.

We show that the Boyle/Hooke law for the pressure/density relation in a gas at constant temperature is a linear equation by showing that increases in pressure always bear the same relationship to the increases in density. We denote the total mass of air inside the pump by m. The density of the air in the pump is $\rho = m/V$, where V is the volume that the air occupies when it is at pressure p. If no air enters or leaves, then the mass will always be m. We first apply a force, F, to the piston, to maintain the pressure, p, in the pump, where the pressure is that force divided by the area, A, of the piston, which is in contact with the air. The gas law tells us that pressure and density are related by the equation $p = \rho RT$, where R is the gas constant and T is the temperature. We now apply a gentle additional push to the piston with force ΔF, so that the new force is $F + \Delta F$. The volume decreases, and the density increases to $\rho + \Delta\rho$. The pressure is now $p + \Delta p$, where Δp can be calculated from the equation $p + \Delta p = (\rho + \Delta\rho)RT$. We obtain the linear relations between force F, pressure p, and density ρ, given by

$$F/A + \Delta F/A = p + \Delta p = \rho RT + \Delta\rho RT.$$

In words, this rule says that the sum of two causes (the original force, F, and the extra force, ΔF, or the original pressure, p, and the extra pressure, Δp) produces the sum of their individual effects (the original density, ρ, and the extra density, $\Delta\rho$). Further, these linear systems also obey a "scaling rule": that is, in terms of cause and effect, if we double the force to $2F$, and thereby double the pressure to $2p$, the effect is that the density increases to 2ρ and the system behaves linearly (this scaling is seen in the above equation when we choose $\Delta F = F$, $\Delta p = p$, and $\Delta\rho = \rho$). Note that if we are impatient and pump quickly, the temperature will increase and we'll feel this warmth if we touch the piston. Then the pressure/density relation is no longer linear because the change of T by an amount ΔT has to be represented in the equations, and this renders them nonlinear.

Exceeding the Force of Any Human Mind

Most of our natural world does not adhere to linearity. The gas law we used earlier is only one of the seven equations describing our atmosphere, and even that equation becomes nonlinear when temperature is allowed to vary simultaneously with pressure and density. When Bjerknes sat down to write his paper in 1904, the remarkable achievements of the astronomers in determining the orbits of planets and comets—and even discovering new ones by analysis of orbital data and further calculation—would have been very much on his mind. Astronomy was a glamorous epitome of the scientific revolution; Bjerknes was determined meteorology should catch up.

One of the most publicized early successes of astronomy (considered as a branch of mathematical physics) was the accurate prediction of the return of Halley's Comet in 1758. This prediction was based purely on calculations using Newton's law of gravity. In 1799 the French mathematician Pierre-Simon Laplace introduced the term *mécanique céleste*, which was adopted to describe the branch of astronomy that studies the motion of celestial bodies under the influence of gravity. Laplace's book on this subject had helped Ferrel teach himself about calculating the movement of objects in the heavens. The emergence of this major discipline within astronomy soon led to other successes: the discovery of the planet Neptune using the equations of celestial mechanics in 1846 as well as the routine prediction of the phases of the Moon, the date and times of eclipses, and the tides. And yet, fundamentally, celestial mechanics involves solving nonlinear problems, so what was the astronomers' secret?

The gravitational force between any two planets is inversely proportional to the square of the distance between them. If their separation doubles, the gravitational force decreases by a factor of four. This is an example of a nonlinear relationship, illustrated in figure 4.4.

When the inverse square law of gravitation is used to compute the motion of one planet orbiting the Sun, the nonlinear problem can in fact be solved exactly. Not all nonlinear problems are insurmountable, and the classical problem Newton first considered is an example of a nonlinear problem that is amenable to explicit and exact solution.

From the outset, Newton realized that his law of gravity was very successful at predicting the motion of the Earth about the Sun, or the Moon

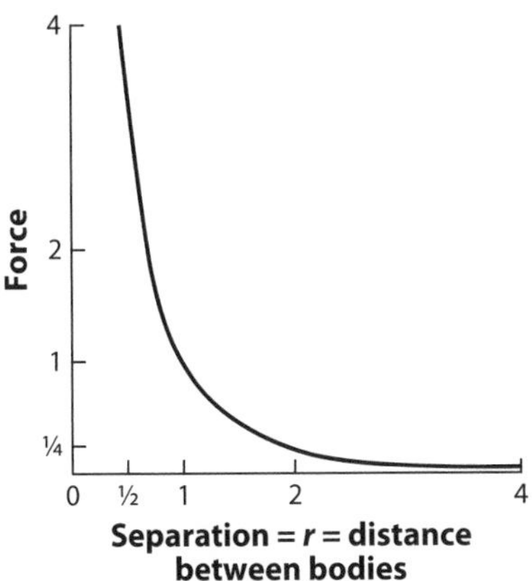

Figure 4.4. The inverse square law of gravity between two objects of mass M and m, separated by the distance, r, is usually written

$$F = (GMm)/r^2,$$

where G is the universal constant of gravity. The curve illustrates that the gravitational force exerted by one body on another decreases as the square of their separation increases and is not a linear relation. Linear relations are readily identified because they show up as straight lines in these graphical representations.

Tech Box 4.1. A Nonlinear Problem We *Can* Solve

To compute the orbit of the Earth around the Sun, we have to solve two sets of nonlinear equations: one for the acceleration of the Earth around the Sun and another for the acceleration of the Sun around the Earth. The latter, however, is practically negligible because the Sun is so much more massive than the Earth that it hardly responds to the Earth's gravitational pull upon it. So we can get a very accurate approximate answer for the orbit of the Earth by neglecting the effect of the Earth on the Sun, treating the Sun as "fixed" in space. This leaves us with one set of nonlinear equations to solve, and it turns out that we can solve them completely. Somewhat remarkably, it is possible to transform the nonlinear equations into linear equations by rewriting them in terms of the reciprocal of the distance from the Earth to the Sun.

The motion of the Earth (of mass m) about the Sun (of mass M) can be found by solving two differential equations:

$$r^2(d\theta/dt) = a \text{ constant},$$

$$m((d^2r/dt^2) - r(d\theta/dt)^2) = -(GMm)/r^2,$$

with r denoting the distance from the Sun to the Earth, and the angle θ giving the Earth's position on its orbit as it travels around the Sun; G is Newton's universal constant of gravitation. In this model problem, we do not consider the daily spin of the Earth (or anything else). (To arrive at these equations we must first prove that the motion lies entirely in a plane, and then (r, θ) are the appropriate coordinates on that plane.)

Next we transform both the independent variable, t, and the dependent variable, r, by using the first equation, which is a mathematical statement of the conservation of rotational momentum of the Earth as it moves around the Sun. That is, we use the first equation to replace derivatives with respect to time, t, by derivatives with respect to θ. We then transform the second nonlinear equation by replacing r with a new variable, $u = 1/r$. We look for a solution where the clock starts at $t = 0$, when the Earth is at its maximum distance, d, from the Sun. We also measure θ from this position, when the Earth is traveling at speed v. Then the transformed equation reads

$$d^2u/d\theta^2 + u = (GM)/(d^2v^2).$$

The differential equation for u is linear, with constant coefficients. This equation can be solved exactly, and the general solution is

$$u = A\cos\theta + B\sin\theta + (GM)/(d^2v^2).$$

We determine the constants of integration A and B from the conditions at $t = 0$. Finally we transform back from u to the radial distance, r. With a little more algebra, the result can be written in the x, y plane, where $r = (x^2 + y^2)^{1/2}$, as

$$(x + ae)^2/a^2 + y^2/(a^2(1-e^2)) = 1.$$

Here $a = d/(1-e)$ and $e = v^2/v_c^2 - 1$, with $v_c^2 = GM/d$. When $0 < e < 1$, this above equation describes an ellipse, which is the orbit of the planet Earth about the Sun.

The original nonlinear problem is transformed into a linear problem that we have solved completely. When we return to the original variable, r, the transformation $r = 1/u$ is a nonlinear relationship, but at this stage of the total calculation the "inverse transformation" is a straightforward exercise in algebra. The key point is that we are able to temporarily remove (via the transformation of variables) the nonlinearity at just the right step, thereby avoiding the necessity of solving a nonlinear differential equation.

about the Earth. In fact, the motion of two fairly equal-sized stars or planets can also be calculated exactly using a more involved set of transformations about a special reference point known as the center of mass. However, Newton also realized that as soon as we introduce even one more planet, then, to use his own words, "the problem [of predicting the subsequent motion], if I am not mistaken, exceeds the force of any human mind." Once again, at the root of the problem lies nonlinearity.

The problem is, if there are three or more bodies interacting with one another under the influence of gravity, then the equations describing their motion become inextricably coupled together because in order to compute how one star or planet moves, we need to know how the other stars or planets will respond to that as yet uncalculated motion. And this is at the heart of the matter, because the motion of one body will affect the other two and vice versa; the motions can "feed back" on one another in extremely complicated ways. The transformation that allows us to transform away the nonlinearity of the two-body problem is no longer sufficient to uncouple the motions of the celestial bodies, and the solution process has to contend with additional nonlinearity. Complete solution is only possible in very special cases.

A perhaps more direct way of appreciating this issue is to consider the dynamics of a system we can construct ourselves. A *double pendulum* can be made by joining two rods together so that the end of one rod is attached to a fixed mounting, such as the edge of a table, while the other end of the same rod is connected to the second rod. This simple mechanical system, which is like letting our lower arm and our upper arm swing freely without friction from both the shoulder and the elbow, can behave quite remarkably.

If there were just one rod, as in a grandfather clock for example, then it would hang vertically when at rest, and if moved to one side and released, it would just swing to and fro in the vertical plane. Gravity acts on the rod, and if there were no friction in the pivot, the rod would

Figure 4.5. A desktop toy based on the principle of the double pendulum shown in figure 4.6.

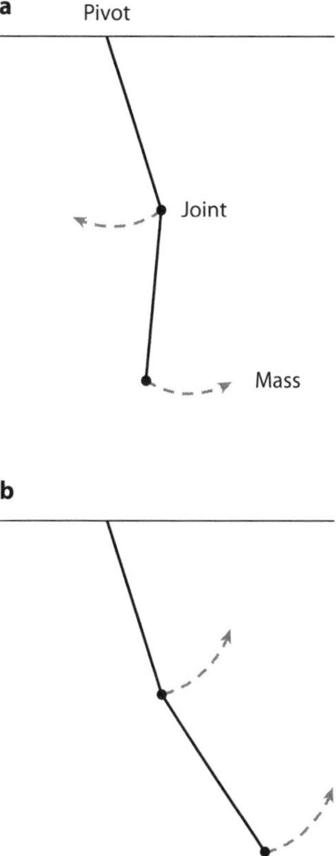

Figure 4.6. This simple illustration of the two-component toy shown in figure 4.5 differs from the earlier two-body problem of celestial mechanics. This model may have unpredictable behavior, as exploited by designers of popular desktop toys. The drawing shows two rods; the first hangs from a table and is joined by a pivot to the second rod. This double pendulum swings freely in a vertical plane. Simple, almost linear, recurring, and predictable motion of the rods occurs when the amplitude of swinging is not very large, as shown here. In figure 4.7 we show unpredictable behavior of this system when nonlinearity becomes more important.

swing back and forth forever. The motion of this single pendulum is completely predictable, and the equations of motion can be solved, or integrated, precisely forever—just as Newton could predict the motion of our planet about the Sun.

When we attach a second rod to the first, the entire situation changes. If we displace the pair from the vertical by just a small amount in the same vertical plane, they move back and forth, either in the same direction or in opposite directions as indicated in figure 4.6. This motion remains regularly recurring and predictable because the interaction between the rods is dominated by linear behavior for all time.

However, if we pull the second rod so that it begins its oscillations from a height greater than the level of the table, then the ensuing motion of both rods becomes highly complicated and will in general become chaotic. Consider releasing the second rod from a point well above the table, and tracing the subsequent motion of its free end. At the moment of release we also impart a motion to the second rod. We then perform two such experiments releasing the second rod at the same position in each but with a slightly differing motion. What would we see when we compare traces of these experiments?

We find that for some time the rods spin around in much the same way. But then the motions become entirely different. We see this in figure 4.7, where the traces start at *S* and separate at *X*. The two experiments then behave in radically different ways. This is the hallmark of chaos.

The cause of this chaotic behavior is feedback. When the two rods are joined together, not only is gravity acting on them, but the motion of one also affects the motion of the other, and vice versa. That is, there is a feedback between each motion of each rod on the other's motion. In this way small differences in the motion of the first rod can initially

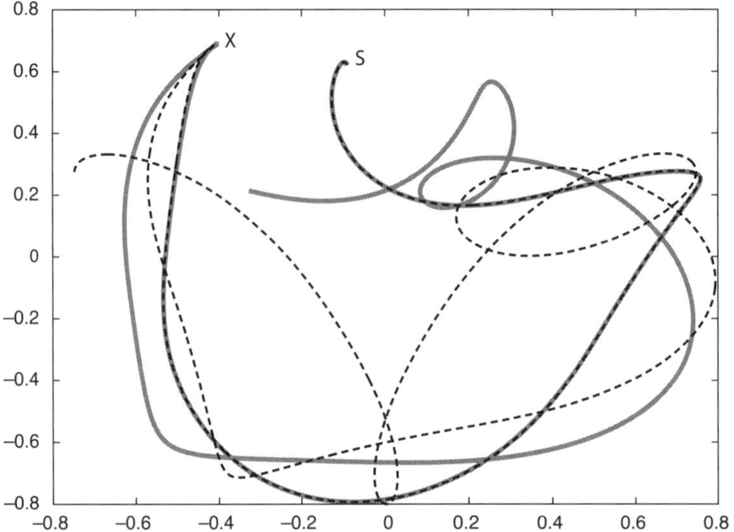

Figure 4.7. The trajectory of the end of the second rod in the experiment of figure 4.6 is shown after it has been released at the point *S* from well above the table. We think of this figure as a picture taken with a long exposure of a source of light attached to the end of the second rod. Then, rather like the postcards that show car lights in cities taken at night, the curves record the flight through space of this point of the double pendulum system. The two curves represent the motion of the end of the second rod after it is released from the same location in space but with slight differences in its initial rate of rotation. The difference is from 40 degrees to 40.1 degrees per second, which is less than 1 percent. Both trajectories start at *S*, near the top/center of the plot at (–0.1, 0.6). The trajectories follow the same path until they reach the point *X* near (–0.4, 0.7) and then they follow very different paths. © Ross Bannister.

make a small difference to the motion of the second rod, but the small differences in motion of the second rod also affect the motion of the first rod (which is feedback). This continues until all the small differences produce very different, and usually chaotic, long-term behavior.

Does feedback always eventually dominate? The answer depends on the situation and the context. As we saw earlier, in the case of small oscillations, sometimes linearity dominates. Stability of the interactions is helped by having one dominant force that the system is governed by: just as the Sun effectively dominates and controls the motions of the main planets about it, similarly, each planet dominates the motions of its moons about it. So the order in the solar system is maintained for very long times in spite of its being a many-body problem.

When will the weather pixels interact like the strongly flipped double pendulum, and when will they behave more like the planets and moons of our solar system? To answer this question we need to take a closer look at the nonlinearity inherent in the fundamental equations that describe weather.

From Pendulums to Parcels

To fly a balloon is to pit oneself against the vagaries of the weather. Steve Fossett, that intrepid American adventurer who disappeared after contact was lost with his plane on September 3, 2007, was a man who courted all the dangers and challenges of the weather in his attempts to fly around the world in a balloon. He became the first person to circumnavigate the Earth by balloon on a solo flight that lasted thirteen days, eight hours, and thirty-three minutes. He took off on June 19, 2002, and covered a distance of 33,195.1 km. If we pause to consider that Armstrong and Aldrin landed on the Moon, a round trip of more than 800,000 km, some thirty-three years before Fossett's successful attempt at the balloon flight, we get an inkling as to the challenge Fossett took on. One of Fossett's colleagues said, "Assuming that all the equipment works, it comes down to meteorological luck."

Even if everything goes right—which experienced balloon pilots acknowledge almost never happens—the technical difficulty of searching out the right winds, staying at the proper altitude, and dealing with variations in temperature and weather make a round-the-world flight an immensely problematic proposition, a game that some pilots compare to three-dimensional chess. The Apollo crews let Newton's law of gravitation do the driving. The point is that the (eventually linear) calculations for the first Apollo mission were capable of being done sufficiently accurately on a modern school calculator. In contrast, a round-the-world balloon trip in our atmosphere challenges modern supercomputers.

If the double pendulum and the three-body problems in general defy exact and predictable solution, what about the behavior of Euler's fluid parcels? To recap from chapter 2, the ingredients of Bjerknes's model are the gas laws and Euler's equations of fluid motion on our rotating planet extended by the thermodynamical laws of heat and moisture. Euler's equations express Newton's second law of motion in terms of the changes to the strength and direction of the wind in response to differences in pressure and density in the atmosphere. Pressure is force per unit area; therefore, the difference in pressure across the faces of a fluid parcel gives rise to a net force acting on that parcel.

We know that pressure in the atmosphere varies from one location to another as the day goes by. Because this variation depends (virtually simultaneously) on the behavior of all the other parcels of air, it turns out that we are dealing with a hugely complicated nonlinear problem. And it is also a subtle problem. To begin to glimpse the nature of this nonlinearity, we first think about a simple experiment: how to compute the changing temperature as we ascend through a still atmosphere in a balloon.

Suppose our flight in a hot air balloon takes place on a calm summer's evening. As the flight progresses we notice a drop in temperature, and this may be due to the combination of two different effects. During the balloon ascent, we experience a decrease in temperature with altitude, and as dusk falls there is also a decrease in temperature due to the setting of the Sun. If we want to account for the change of air temperature in the basket of the balloon over a period of time during our flight, we have to consider both of these effects. For simplicity, we ignore horizontal motion of either the balloon or the air, as shown in figure 4.8.

To plot a graph of the change of temperature with time during our flight, we then need to combine three pieces of information: (1) the change in air temperature due to the setting of the Sun, which is the measure of the change of temperature as time passes at a particular location; (2) the change in air temperature by varying our height in the atmosphere, which is a temperature change with respect to position; and

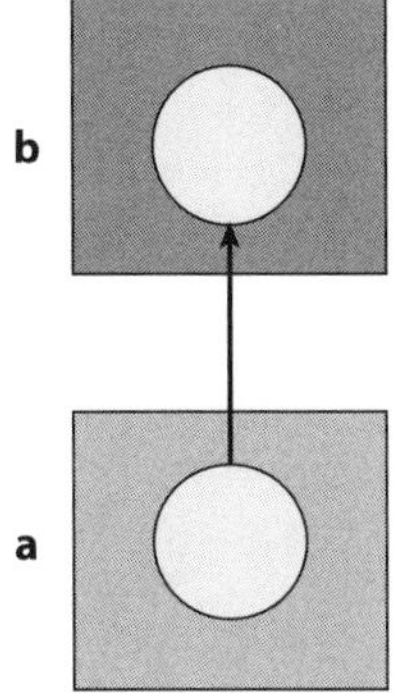

Figure 4.8. By the time the balloon reaches position b, which is vertically above a, the temperature of the surrounding air is cooler both because it is later in the evening and because the balloon's altitude is greater than at a; (usually) the atmosphere is cooler the higher we ascend, as it was on this flight.

(3) the rate of our ascent—the faster we ascend, the more rapid the total change in our temperature will be as we combine the first two effects. If our balloon is tethered, then we only measure the first contribution. If we tether balloons at two different heights, the difference in the measured temperatures at any fixed time would measure the second contribution. But the change in temperature in our balloon as we change altitude is dependent on both 1 and 2 as well as the time it takes us to ascend.

So the expression for the rate of change in temperature we experience as we travel up with the balloon reads

Rate of change of our temperature = rate of change of temperature at current location + rate of ascent × rate of change in temperature with altitude.

In terms of the notation introduced in the tech boxes of chapters 2 and 3, using the total derivative D/Dt for the rate of change measured by a thermometer traveling upward with us at speed w, this expression is

$$DT/Dt = \partial T/\partial t + w(\partial T/\partial z),$$

where T is temperature, t is time, and z denotes our altitude.

This formula for the rate of change of temperature that we experience (as we fly in our balloon) is readily understood. We add the rate at which the environment is cooling at our location to the result of multiplying our rate of ascent by the rate of variation of temperature with altitude. This gives us the numbers we require when we use the same units (such as miles or kilometers) to measure things. We now make some assumptions to simplify the arguments, without losing the salient features. Assume that the decrease in air temperature at any altitude

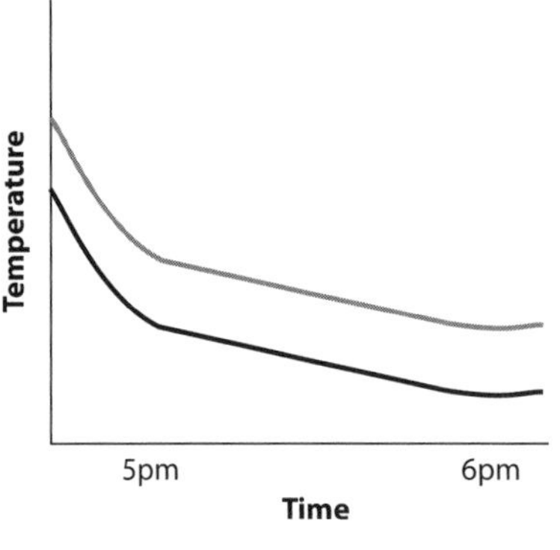

Figure 4.9. Each curve shows the decrease in temperature at a specific altitude with time. The lower (colder) curve shows the temperature decrease at a higher altitude. Between 5:00 and 6:00 p.m., the temperature decreases uniformly by 5°C at all altitudes the balloon travels through.

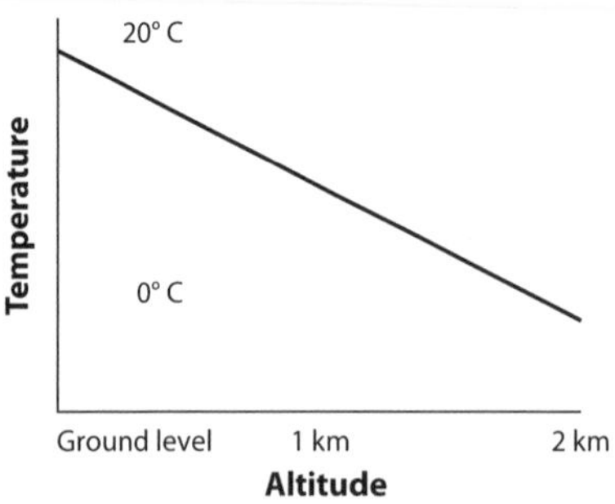

Figure 4.10. This descending line shows the decrease in temperature from 20°C to 0°C over a 2 km increase in altitude, at 5:00 p.m. (the slope is $\partial T/\partial z = -10$°C/km).

follows the same pattern as the Sun sets, as shown in figure 4.9. Referring to figure 4.10, we also take the decrease in temperature with height to be a straight line (a linear relationship); the product of a constant rate of ascent with this relationship produces another linear relationship but with a different slope for the line. Finally, add the cooling effect due to the setting of the Sun, shown in figure 4.9, to this product, thus getting the final linear relationship, shown in figure 4.11, for the total decrease in our temperature to −5°C at an altitude of 2 km at 6 p.m.

This all seems straightforward enough, and the calculations are slightly lengthy but linear. However, in practice, balloons do not usually ascend at a constant rate. If we are not using the burners, our balloon ascends freely at a rate that depends on its buoyancy. This buoyancy is dependent on the temperature difference between the air in the balloon and that of the surrounding atmosphere. This means that the last term in our formula—rate of ascent × rate of change in temperature with altitude—is the product of two terms that are not independent of each other. Further, the local rate of change of temperature might itself vary with height, and temperatures at different heights might cool differently as the evening passes.

This mutual dependence brings about a more difficult scenario for analysis; fortunately, pilots know in practice what to do, so we enjoy our flight. The changing flight pattern of our balloon means that we travel through a different environment, and this change of our environment further changes our flight pattern, and so on. This is nonlinear feedback, and it potentially affects every calculation of air movement that we make. In practice, the pilot uses his or her judgment to operate the burners to compensate for these changes, and all is well, provided the changes are not too large—an advantage of flying on calm evenings.

We complete our balloon story a little more analytically. When the rate of ascent depends on the difference between the fixed temperature of the balloon (which is assumed to be perfectly insulated) and the temperature of the environment, then the rate of ascent will increase with time. This is because the environment is cooler as we get higher, and buoyancy depends on the temperature difference. Instead of the linear relationship depicted in figure 4.11, we now find that the change of temperature with time is given by a nonlinear relationship (no longer a straight line), as depicted schematically in figure 4.12. When this situation persists, the balloon will accelerate to the top of the atmosphere, providing it does not burst. In practice, the much cooler surrounding air cools the gas in

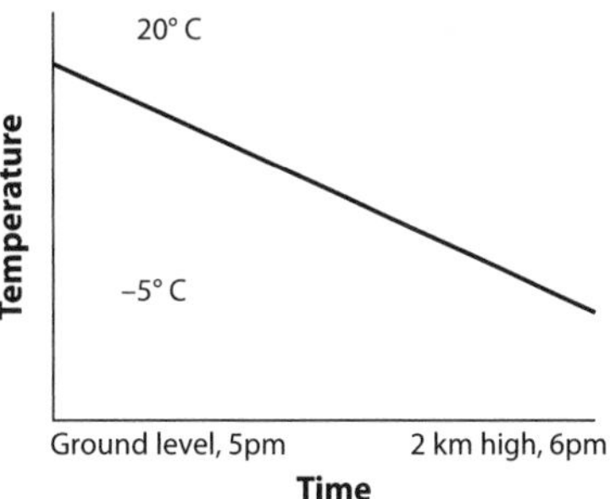

Figure 4.11. This straight line shows the rate of change of temperature experienced by the balloon passenger as the balloon ascends at constant speed. It is the product of the rate of ascent, say 2 km of height per hour, with the relationship in figure 4.10, which is −20°C in the hour, plus the cooling due to the setting Sun shown in figure 4.9, which is −5°C in the hour. Again we have a linear relation but with slope $DT/Dt = -25$°C/hour.

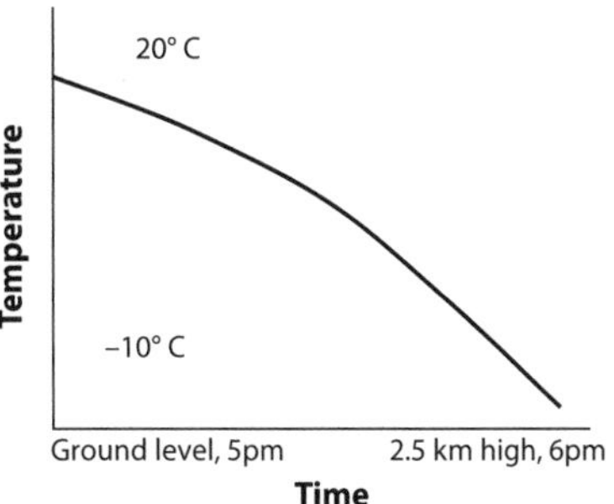

Figure 4.12. This shows the rate of change of temperature as the balloon ascends at a rate depending on the temperature of the air that the balloon travels through, assuming the air in the balloon does not cool down. The slope of the curve—that is, the rate of change of temperature with elapsed time—varies with time, which is characteristic of a nonlinear function.

the balloon, and some of the gas escapes, so we eventually start sinking unless the pilot lights the burners.

The discussion about balloon flights has shown us a little of what is involved in calculating and understanding the parcel-based derivative *D*/*Dt*. In usual flying conditions, a wind blows the balloon horizontally as well. But what if the wind transporting the balloon depends on the state of the transported air parcel? Air parcels heated at the ground ascend in thermal plumes, just like the flow of air up Bjerknes's heated chimney in figure 1.6. When the vertically transported air parcel contains significant water vapor, condensation eventually occurs and cloud forms. The more vigorous this process, the bigger the cloud, as figure 4.13 suggests. But when cloud forms, the latent heat of condensation is released back into the air parcel, which further affects its buoyancy. We see the appearance of this nonlinear feedback whenever we see convective clouds developing. However, as we discuss in the next section, even a dry wind at a constant temperature has the capacity to develop nonlinear feedback.

Figure 4.13. Cumulus clouds are shown as they are carried by the wind. In describing this growing cumulus cloud, it is helpful to think of each small cloud as marking an air parcel that floats at its neutrally buoyant level, a little like a balloon. But what happens if the cloud also affects the wind carrying it?—this is a feedback process, a key example of nonlinear behavior. The creation of spectacular cloud formations is due in part to the "invisible" nonlinearity of *D*/*Dt*. © Robert Hine.

Beyond a Million-Dollar Challenge

As the balloon was carried by the wind in our previous discussion, we now think of an air parcel transported by the airflow. We apply a similar reasoning to that of our earlier computation of the change in temperature, except that w, the vertical motion of the fluid parcel, will be governed by Euler's equations. We focus first on the vertical motion. Here the vertical acceleration will be proportional to both the change in pressure with altitude and to the buoyancy of our air parcel. The rate at which the heated parcel is accelerated upward by buoyancy is constant only if the temperature and density are constant. But we know from the gas law that pressure, density, and temperature are dependent on one another (as we saw in chapter 2).

This means that calculating how the temperature of an air parcel changes involves a subtle interplay between the gas law and Euler's equations. This interplay can lead to very complicated behavior because the rate of ascent and the temperature of the parcel can modify each other, and it is usually far from straightforward to solve such problems using the standard methods of calculus. This interplay reveals itself in the product of the last two terms in our formula for D/Dt following figure 4.8, and when the two terms are dependent on one another, we have a nonlinear relationship. In fact, such feedback processes are usually the cause of many severe weather events. Furthermore, the presence of moisture complicates the situation even more. When the parcel of air is humid, cooling leads to moisture condensation, cloud forms (and rain may even occur) just like heavy dew on the ground during a cool summer's night. Condensation of water vapor releases yet more heat into the surrounding air, so we have to go through the gas law calculation all over again. This latent heat release from condensation is a bit like our pilot lighting the burners to reheat our balloon—usually the balloon then rises. Thunderstorms rely on this feedback process to provide their energy and to grow their most spectacular cumulonimbus anvil-headed clouds up to heights of more than ten kilometers.

In the previous paragraph we focused on the vertical motion only. So next we find a formula for expressing the rate of change of wind speed and direction (that is, the acceleration) of the air parcel with time. Measuring the rate of change in the wind velocity, that is the acceleration of the air parcel itself, will have to account for both the change in the strength and direction of the wind at a particular location and the change in strength and direction of the wind from one location to another.

Let us go through this step by step. The direction and speed of the wind may be changing at a particular location (which is what we measure with a weather vane and anemometer, respectively). This accounts for the first term after the equals sign in the equation in the box below. However, we also have to account for change in the wind as we follow the air parcel, as we indicate in figure 4.14, because conditions may change from one location to another, just like the temperature varies with altitude. The rate of change of wind velocity of the air parcel as it changes location is therefore given by the rate of change in the wind from one location to the next multiplied by the rate at which we move from that location.

So the acceleration of the wind is the sum of two contributions:

Total acceleration of the flow = acceleration at current location
+ wind speed × rate of change of velocity of the flow
in the direction of air parcel motion.

Note again that the last term is the product of two terms that depend on the wind speed and direction and are therefore dependent on one another, and the expression is nonlinear. So the change in the wind velocity depends on the wind itself. In other words, we have to account for how the moving parcel that makes up the wind actually blows, or "carries," the air parcel itself. This amounts to calculating how much the

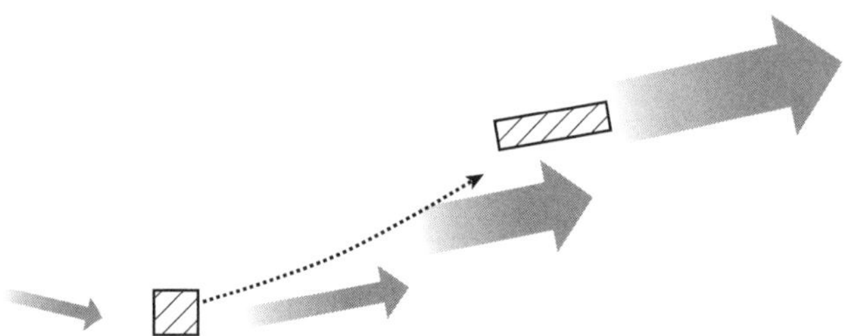

Figure 4.14. In this illustration a hatched rectangular air parcel is moving in a wind (like a balloon or a small cloud), and the strength of the flow is shown by the relative sizes of the arrows. The amount by which the air parcel will be accelerated depends on the acceleration of the flow at its current location, and on the rate at which the flow is accelerating along its trajectory (denoted by the dashed arrow). Our air parcel may be stretched and lifted as it moves, which may complicate the situation further.

"wind blows the wind"! Occasionally this results in a "stampede," and then the wind accelerates, almost out of control, creating storm-force winds, hurricanes, and tornados, depending on the environment. The feedback in large storms is typically "fed" from the latent energy of the condensing humid moist air being transported by the very wind itself.

We need to know how much of the acceleration is due to the change in the speed and direction of the parcel at a particular location, and how much is due to the motion of the parcel in a particular direction. This last statement uncovers the devil in the detail: in order to compute the acceleration of the parcel in a particular direction, we rely on knowing the path the parcel will follow—in other words, its motion. But the motion is unknown until we solve the equations. And it gets worse. The pressure gradient force itself depends on the motion of the air—that is, the wind we are trying to calculate—and so does the density. And the density is affected by the air temperature, which itself is affected by any moisture condensation. Feedback upon feedback upon feedback. These nonlinearities conspire in differing environments to cause many, many wildly differing winds. Our weather equations relating our seven weather pixel variables are all coupled together, so we can imagine that the temperature, the wind, and especially the moisture can feedback on each other in a myriad of subtle ways, as indicated in figure 4.15. To make progress, we need to untangle this Gordian knot.

The outcome of applying Newton's second law of motion to a fluid parcel is that the actual wind at any instant will influence the change in that very wind itself. Feedback processes, as we have already described in terms of the double pendulum, can place severe limitations on the prediction of future events.

Euler's paper describing the equations of ideal fluid motion (with no heat, density, frictional, or moisture processes, and no Coriolis effect) was published in 1757; it was the heyday for applying methods of calculus and analysis to problems in physics. Such was the pace of progress in solving other problems that after five years Euler was dismayed that no one had been able to solve his equations for fluid motion.

In 2000, nearly a quarter of a millennium after the equations were published, a prize of one million dollars was offered to anyone who could prove that, in the absence of any restrictive assumptions, solutions to equations such as Euler's always actually exist. We know that particular solutions exist, and we solve these types of equations using approximate

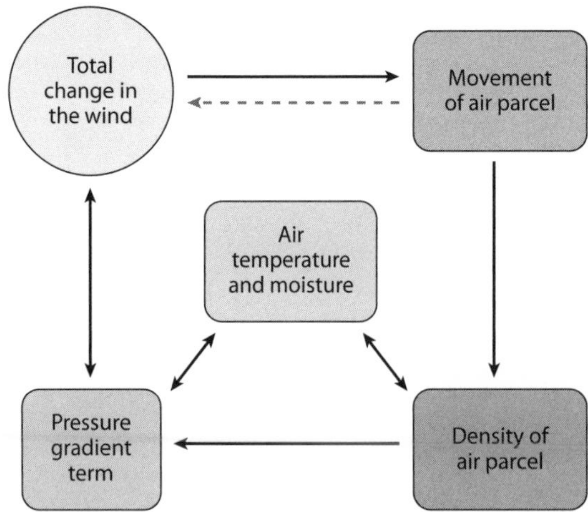

Figure 4.15. This diagram schematically illustrates how the force acting on a parcel of fluid (the pressure gradient term) helps to cause a change in the velocity of that parcel; in turn, the changing winds then affect the movement of each air parcel. But this air parcel movement may produce a feedback, illustrated by the dashed arrow, that modifies the wind itself and modifies the pressure gradient term. The motion is further modified through the pressure gradient term by the change of the air density, which itself is linked to the air movement. The air temperature and moisture are also linked to the density and pressure, so providing many other possibilities for feedback. At different times one mechanism will dominate, and this usually produces distinctive weather. But the questions are, which mechanism, when, and where?

numerical methods every time we compute a weather forecast, but this is not the same as proving that solutions will exist in general, and forever. Success with this proof will bring more than a substantial financial reward: the winner will without question earn a place in the history books of mathematics and physics. The prize remains unclaimed to date, and the intractability of the problem lies in its nonlinearity—the feedback process as the wind blows and modifies the wind itself.

Nonlinearity and feedback lie at the heart of the evolution of our atmosphere and make our equations essentially impossible to solve explicitly and analytically; algorithmic techniques and supercomputers are needed to forecast the weather, even approximately—a pertinent reminder of the far-reaching consequences of Richardson's ideas.

These difficulties involving feedbacks make us appreciate why the circulation and vorticity theorems we mentioned in the previous chapters are so valuable. They tell us under what conditions vortex motion

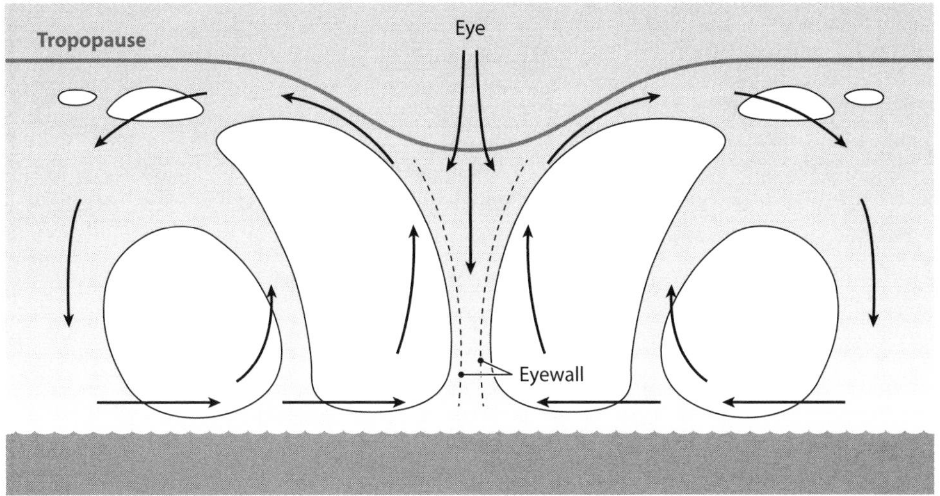

Figure 4.16. Hurricanes form when a positive feedback is established between the heat energy from the condensation of the rising water vapor and the storm winds (see figure 3.8). The transfer between heat energy and the mechanical energy of the winds is related to that of a giant steam engine (see figure 2.14). The tropopause marks the boundary between the troposphere and the stratosphere; the hurricane is moving across a warm tropical sea, which supplies the water vapor. The arrows in the middle of the figure show the descending clear air inside the thick cloud "eyewall."

will persist, or under what conditions it might be amplified. As figure 3.8 shows, a hurricane has vortex-like motion. Why does such a hurricane persist so destructively in spite of the local chaos of the gusting winds and all the deluges? The link between circulation and the heat and moisture processes helps to explain this (see figure 4.16). The power and utility of the circulation theorems lies in the fact that they describe the evolution of a key variable, the vorticity, which is often a signature of large-scale weather patterns.

Today supercomputers are used to solve the basic equations, much as Richardson himself had calculated, with sufficient accuracy both to enable the study of a broad spectrum of nonlinear phenomena and to forecast the weather usually very reliably over most parts of the Earth, even if only for a few days. But despite the power and the increasing reliability of computer simulations, we cannot escape the fact that weather forecasting is firmly based on a set of equations that we really know very little about. In Bjerknes's day we knew how to compute and predict the orbits of planets; now, using Einstein's theory of gravity, we are able to formulate theories and compute models of the beginning and the end of the universe. By many measures, we know far more about the solutions to Einstein's equations of general relativity, which describe the large-scale structure of the universe, than we know about solutions to Euler's equations of fluid mechanics. In fact, we know even less about the equations for weather forecasting: there is just so much detail, complicated by so many layers of feedbacks.

The next great advances in meteorology would be made when ways could be found for analyzing large-scale atmospheric flows so that the governing mechanisms were isolated and then described in terms of simpler models and equations. Naturally, ideas to simplify the real behavior were needed, and there are many ways of simplifying the equations of motion. The Bjerknes circulation theorem is just one way forward. Thousands of research papers have been devoted to this task, but whatever technique is used, we need to come to terms with the grasp of nonlinearity, and not just that involved when the wind blows the wind.

INTERLUDE

A Gordian Knot

In the first half of this book we set out a way to describe weather using a computer. We introduced "weather pixels," which make up a hologram of our planet's weather. The hologram shows the wind, warmth, cloud, and rain at each fixed location in our atmosphere and at a given time. Then, just as a movie advances a sequence of images, we need to advance our weather pixels in time, and this requires rules to relate each pixel both to its neighbors and to earlier pixels. But if the weather pixels are to follow actual planet Earth weather, as shown in figure I.1, then the rules for changing each weather pixel need to

Figure I.1. Earth and Jupiter. The flows and patterns in these pictures are worlds apart yet have many features in common. These flows are described by the same basic mathematical equations as those at the end of chapter 2. Vorticity is ubiquitous in the atmospheres of both planets, but the detailed motion is different. The challenge in simulating these fluid systems using computer models is to design software that will represent and respect such differences even though the code is based on the same equations. Photos courtesy of NASA.

encapsulate our knowledge of the physics of the atmosphere in just the right way.

Having got the rules to advance the weather pixels into the future, why can't we just let the supercomputers work tirelessly away, much as Richardson had done, and let them calculate the next ten years' weather? The answer, in a word, is feedback. Feedback loops operate in the weather and sometimes lead to storms that can range in size from tornadoes to thunderstorms to tropical hurricanes.

The problem with rule-based reductionism to the simplest arithmetical models for solution by computation, as carried out by Richardson, is that our choice of pixel size and the time step will always limit the extent to which we can resolve the various physical processes at work. Feedback between cooling, moisture release, and the blowing winds occurs to some extent throughout the atmosphere all the time. Having chosen our pixels and fundamental rules, we must guide the computation so that it follows what really happens, not an accidental artifact. This is why we need to identify overarching holistic principles that the rules must satisfy. Bjerknes's circulation theorem is one such element that helps us simulate actual weather while avoiding that which is imaginatively possible.

Our first task in the second half of the book is to cut the tangled knot of cause and effect, tied sensitively but robustly by various types of nonlinear feedback—when the effect actually modifies the cause. Chapters 5 and 6 describe the way that this thinking led to the first successful but highly simplified weather prediction via computer, carried out in 1950 by a team in Princeton. Chapter 7 looks to the future and identifies the underlying holistic principles with, somewhat surprisingly, geometry, so that we might improve our rules for pixel evolution in the abstract weather life-history spaces of our computer models.

Finally, in chapter 8 we evaluate the present state of play. Even with the most optimistic assumptions on computer and satellite development over the next two decades, we will not be able to quantify the impact of all the planes, trains, automobiles, and buildings—let alone butterflies—on the atmosphere. We will not know or be able to predict all the finer-scale interactions and feedbacks in weather. So we need to refine our ability to forecast the main features of planet Earth weather in the ever-present unknown of the detail. Understanding the mathematical basis of the weather model helps us improve both the weather pixel description and the weather rules for pixel advancement into the future. Success with this program—a contemporary adaptation of Bjerknes's vision—will increase the confidence with which computer predictions of weather and climate can be made over this century.

FIVE

Constraining the Possibilities

The quantitative model envisaged by Bjerknes and created by Richardson is considered a "bottom-up" view of weather. It includes as much detail as possible and focuses on how each local region of air influences and interacts with its neighborhood. By writing down the laws that govern the detailed physics of the forces and winds, the heat and moisture, we can proceed to simulate "weather"—the consequence of these complex interactions. The problem with this reductionism, à la Descartes, is that the many component parts can interact in hugely complicated ways.

The Bergen School identified the salient features of certain weather systems and assimilated their ideas into qualitative models. The key mechanisms behind familiar weather patterns had been isolated from the infinity of possible phenomena allowed by the governing equations, but no mathematical description that directly related such ubiquitous phenomena to the equations was available. By initiating a program to quantify one of the conceptual models that had emerged from the Bergen School, a young Swedish meteorologist was to create the first mathematical description of large-scale recurring weather. This chapter is about his remarkable life's work—every bit as important, if so totally different, from that of Bjerknes.

In Pursuit of Bjerknes's Vision

Lieutenant Clarence LeRoy Meisinger ascended in a balloon from Fort Omaha, Nebraska, on March 14, 1919, and reported on his trip in the

U.S. Weather Bureau's *Monthly Weather Review* a month later. An artistic and talented writer, Meisinger was born in 1895 in Plattsmouth, Nebraska, graduating with a degree in astronomy from the University of Nebraska in May 1917, just after the United States entered World War I. He joined the army in June 1917 and played the French horn in the 134th infantry band. In April 1918 Meisinger was transferred to the new Meteorological Service of the Signal Corps. Soon he was appointed chief weather officer at Fort Omaha, where the army had its Balloon Training School, and there Meisinger gained his balloon pilot license.

So began Meisinger's love affair with the upper air: his devotion was evident from a report: "Below the balloon was a gently undulating sea of fog, soft as down, and delicately tinted as mother-of-pearl," and overhead floated "a layer of alto-cumuli opaled by the rising sun, and ever varying in iridescent splendour. . . . This journey . . . enabled us to actually penetrate and become part of the wind circulation of a strong cyclone."

With war training at an end, Meisinger next joined the U.S. Weather Bureau in Washington in September 1919, where part of his duty as a scientist was to help edit the Bureau's *Monthly Weather Review*. He enrolled in graduate school and in 1922 earned his doctoral degree with a thesis entitled "The Preparation and Significance of Free-Air Pressure Maps for the Central and Eastern United States" (see figure 5.1). Meisinger's devotion to aviation, with its ballooning, dirigibles, and growing numbers of aircraft, made him realize that understanding the behavior of the "upper air" was necessary to protect pilots and passengers during bad weather, and this was precisely where observations were least available. After publishing his thesis in the *Review*, Meisinger was on his way to being recognized as the leading aeronautical meteorologist in the United States.

By the spring of 1924, Meisinger and Lieutenant James T. Neely, his pilot, both skilled and experienced balloonists, were engaged on a series of balloon flights from Scott Field, in southern Illinois. Their intention was to explore the air motion within cyclones. The Bergen scientists were acquiring a detailed understanding of these weather systems from systematic observations mainly at sea level, and it was time to study them aloft. However, these April and May months were unusually difficult weatherwise, and their last flight to collect data was on June 2.

As the Sun set, the cooling on the balloon caused it to descend to almost 300 meters; they threw away ballast and the balloon rose to more

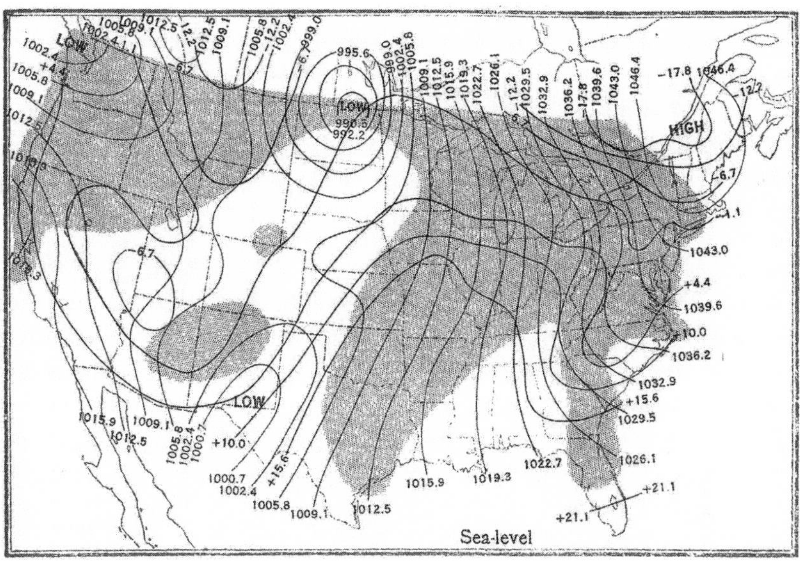

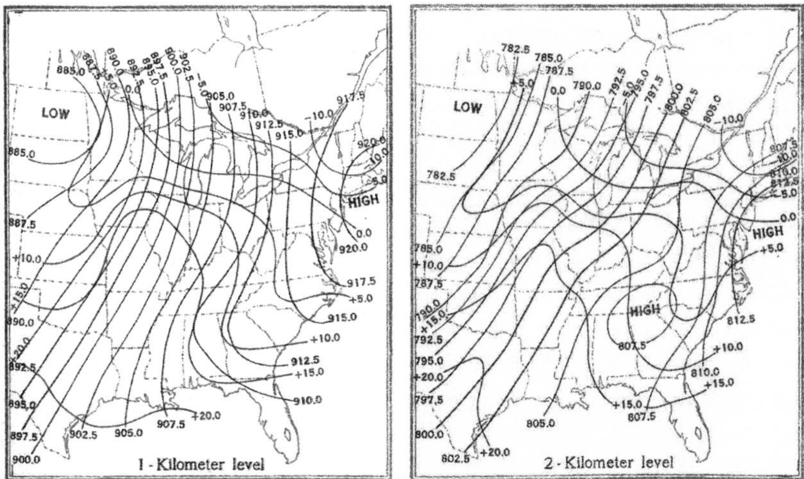

Figure 5.1. Meisinger's early work catalogued the paths of cyclones crossing the Midwest. Meisinger realized flying was the future, and scientists needed to understand the upper air motion where flight would occur. This figure is taken from Meisinger's paper, "The Preparation and Significance of Free-Air Pressure Maps for the Central and Eastern United States," published in *Monthly Weather Review* 50 (1922): 453–68. © American Meteorological Society. Reprinted with permission.

than 1,500 meters only to begin descending again rapidly. Again ballast was released at 300 meters, and again they ascended, this time to nearly 1,800 meters. Again the whole procedure of rapid descent, ballast release, and climbing—this time to above 2,200 meters—was repeated in the unstable dangerous air; yet again this was followed by a rapid descent, but this time hitting the ground near Milmine, Illinois. The crash released more sand ballast. As the balloon climbed once more, they were hit by a lightning bolt, which killed Meisinger. Neely attempted to parachute from the burning balloon but fell to his death.

From the outset, both men would have appreciated the risks associated with their venture, but what drove Meisinger to put his life on the line in this way? The evidence suggests that he was totally committed to furthering the science encapsulated in Bjerknes's vision. The professional forecasters' resistance to embracing the new ideas that were emerging from the Bergen School dismayed Meisinger as well as Bjerknes. The seminal papers from Bergen describing the critical features of midlatitude storms were published in *Monthly Weather Review* in February 1919, and American forecasters more or less immediately discounted their ideas. However, a year later Meisinger published his own paper in the *Review* in which he noted the striking similarities between the Norwegian cyclone model and a severe storm that had crossed the United States. His advocacy of ballooning as a means of furthering meteorological science was fueled when, in 1920, a balloon race in Alabama was won by a Belgian meteorologist who had managed to position his balloon so that he gained speed from the strong winds above the steering line, or warm front, of an advancing cyclone, just as described by the Bergen School (and shown in figure 3.9). Meisinger's devotion and contribution to his subject is now honored each year by the American Meteorological Society, which presents the Clarence LeRoy Meisinger award to outstanding young meteorologists.

By the late 1920s, Vilhelm Bjerknes would have been justifiably proud of the success of the Bergen School, but it is clear from various letters and correspondence that he was also disappointed because his original manifesto for forecasting, based on the direct use of the equations of physics, had undeniably foundered. How can weather patterns be seen to evolve through the complex interconnected processes encapsulated in the seven basic equations? The breakthrough came in the 1930s, when a young mathematician-turned-meteorologist found a way to unleash the predictive power of Bjerknes's theorem. And, like Meisinger, this young scientist's formative training was inspired by Bjerknes's vision, seen firsthand when he spent nearly two years at the Bergen School.

Carl-Gustaf Rossby was born in Stockholm in 1898, the year in which our story began. Rossby specialized in mathematical physics at the Stockholm högskola, graduating in 1918. He next went to the Bergen School, watching the teams' methods and working in the presence of the master (see figure 3.6). Then he went to Leipzig in the summer of 1920 to the Geophysical Institute, also founded by Bjerknes. In the summer of 1922 Rossby visited the Prussian aeronautical observatory at Lindenberg, near Berlin, before returning to Stockholm to spend 1922–25 training at the Swedish Meteorological Institute. During the same period he wrote a thesis with the renowned mathematician Erik Ivar Fredholm. Rossby even found time to travel through the east Greenland pack ice on the oceanographic vessel *Conrad Holmboe* in 1923. Rossby would continue this frenetic lifestyle until his untimely death in 1957.

In 1925 Rossby left for the United States, where he remained for twenty-five years. During this quarter century Rossby revolutionized meteorology in the United States both practically and theoretically. Despite being very bright, inquisitive, active, and extremely well trained, it would take Rossby the next seven years to absorb and interpret the Bergen School ideas in his own mathematical language, and it was the mid-1930s before Rossby realized his own view that the cyclone weather systems had more to do with the westerly winds in the upper part of the troposphere than the Bergen polar front. But having Bergen on his curriculum vitae was clearly very beneficial following his move to the United States.

At first, Rossby was given a desk in the furthest corner of the U.S. Weather Bureau, which effectively put up an impenetrable barrier to the young Swede's exuberant schemes, according to Horace R. Byers. Byers, born in 1906, spent much of the period between 1928 and 1948 helping to carry out Rossby's ambitious plans, initially in the commercial world of flying and eventually at the universities of MIT and Chicago. It happened that others realized Rossby's great potential; already by 1927 the Guggenheim Fund was supporting him, as was the U.S. Navy via Francis W. Reichelderfer.

Born in 1895, Reichelderfer knew about weather from the practical side as a navy pilot of a biplane, and he realized (following a lucky escape in December 1919) that understanding weather could make the difference between life and death for a pilot. During a major series of cross-country flights to demonstrate the usefulness of aviation, the famous pilot Richard Byrd commented favorably on Rossby's forecasting abilities, having scorned those of the Weather Bureau. Following this Rossby was declared persona non grata in the Weather Bureau. On the back of his success in aviation forecasting, Rossby was invited to go to California where he set up the first weather service for a trial airline flying between Los Angeles and San Francisco. In 1928 the Guggenheim-funded Department of Meteorology at the Massachusetts Institute of Technology (MIT) asked him to launch a new graduate course, in the beginning taken by four Navy officers.

So, by his early thirties Rossby was accepted and highly regarded by the majority of his colleagues, and increasingly by the authorities as the decade wore on. In 1939, newly naturalized as an American citizen, Rossby was asked to be assistant chief of research and development at the U.S. Weather Bureau, under their new chief, Reichelderfer, and together they modernized the bureau. In 1941 Rossby came to Chicago to head the Meteorological Institute at the University of Chicago, and, although based there for ten years—a long time for him—he continued

Figure 5.2. Carl-Gustaf Rossby 1898–1957. In the preface to the "Rossby Memorial Volume" (which was originally intended as a celebration to mark his sixtieth birthday), his colleagues wrote, "His death, at the age of 58, deprived the world of science and his many friends of a man who, by force and charm of personality, power of intellect, and indomitable spirit and energy was a central figure in the explosive development of meteorology that has taken place during the last three decades." Courtesy MIT Museum.

to make many trips, both at home and abroad. Rossby attracted many leading scientists to the department and strongly influenced their research, so that some authors refer to this period as the Chicago School.

Rossby became president of the American Meteorological Society (AMS) and helped establish a number of scientific journals that today are among the most widely read by professional meteorologists. While president of the AMS, he gave the society new goals to pursue, including supporting cooperation between government and private weather companies, encouraging research in "economic meteorology," and promoting education in meteorology. The international meteorological community continues to aspire to these goals today.

Rossby later became one of the first to speak out about the dangers of atmospheric pollution and acid rain, and he was featured on the front cover of *Time* magazine in 1956, which carried an article about the emerging issue of air pollution and the environment. But all of these achievements are independent of the massive contribution he made to the basic science itself. Those early days at the Bergen School, where he was surrounded by brilliant research, had really paid off.

Historians have noted that Rossby had striking character attributes that contributed to his success as an all-rounder, capable of thought and action in both the commercial world and academia. His organizational abilities, excellent communication skills, and exceptional judgment made him an inspirational leader, but he also possessed insight and ingenuity, which made him a scholarly research scientist capable of developing new forecasting methods.

The scientist and leader was therefore both a theoretician and practitioner: his natural abilities as a mathematical physicist and a practical meteorologist equipped him with just the right skills to move such a diverse subject as meteorology forward. Despite his training in the qualitative and graphical methods of the Bergen School, in many ways he was like Richardson—he wanted to get practical benefits out of the mathematics alone and, like Bjerknes too, he knew of the formidable obstacles to doing just that. He was intrigued by the puzzle: how were the conceptual models of the Bergen team related to the fundamental equations, and how could we access useful information hidden in these arcane mathematical laws?

To facilitate a mathematical approach to understanding atmospheric motion, it was essential to find simpler quantitative descriptions of the evolving large-scale patterns in the weather that the Bergen scientists had identified so clearly. The fundamental equations represent many exceptionally complex interactions and feedbacks. But the plethora of detailed, local physical interactions invariably produces a relatively small number of recurring larger-scale phenomena, such as the fronts and anticyclones that we typically associate with daily weather in the middle latitudes. These organized, coherent large-scale structures, dwarfing local mists and breezes as they drift across our continents and oceans day by day, represent a small subset of the uncountable number of possible fluid motions that could be produced by the myriad highly local interactions at the smallest scales (see figures 3.12 and CI.7). So although the qualitative features of "weather" were becoming better understood by the Bergen School, an explanation of the ubiquity of such features was not apparent from an examination of the equations.

It turns out that we can learn a great deal about the patterns that broadly recur every few days by studying only a few controlling mechanisms. These can be quantified in terms of relatively simple mathematical models. Simple models had already solved the problem of how to make predictions of the major tides in our oceans. The main behavior—that of high and low tides and their approximate rise and fall—is calculated from a model of the interactions between the Earth, Sun, and Moon. All detailed wave motions arising from the winds or sloshing around on coastlines are completely ignored, as are many other factors. Then local rules based on decades of observations are used to calculate, say, the delay of high tide as the water rushes around islands and through channels on local coastlines. Exceptionally high or low tides are governed by certain alignments of the Earth, Moon and Sun. Storm surges on high tides may bring significant damage, so weather forecasting to determine low pressures brings in further complication. But all this builds on the base of the simplified Earth, Moon, and Sun system. Now, in a similar way to predicting tides, we must isolate the major factors that constrain the behavior of our atmosphere.

By the early 1930s Rossby was set to embark on his quest to find the key to understanding and predicting the larger-scale motion of cyclones, fronts, and entire air masses amid the complex nonlinear mathematics of air motion in the presence of heat and moisture. Nothing like this had been attempted before in meteorological science, but the goal was clear, and Rossby posed the question in the early 1930s: "How can the general equations of hydrodynamics, formulated more than a century before, be adapted to give a self-consistent and mathematically tractable description of the large-scale motions of the atmosphere?" Rossby paved the way to an answer, and in the next chapter we describe how one of the people he inspired came up with the full breakthrough. But before we describe these events, we explain that the key to tackling this problem is to first identify, as a consequence of the structure of the equations themselves, the quantities remain constant throughout the detailed interactions.

Conserving Order

There is one striking feature of all physical laws that plays a crucial role in Rossby's work. In *The Character of Physical Law*, the American physicist Richard Feynman devotes an entire chapter to principles that emerge as important consequences of the laws of physics. The principles involved are called *conservation laws*, and, in fact, they have already underpinned much of our story. They were central to Halley's and Hadley's theories; they lie at the heart of von Helmholtz's and Kelvin's theorems of vorticity and circulation for a perfect fluid, and they are also fundamental in explaining the role of energy. And of course the circulation theorem of Bjerknes is another example.

So what does a conservation law actually mean? In practical applications it means that a quantity or a number can be calculated from measurements, and then the quantity can be calculated again as nature undergoes a multitude of changes; we find its numerical value remains the same. That is, its value will not change; it is "conserved" in spite of the many minor, and possibly some major, things that might happen during the process.

We first illustrate the idea with the same example that Feynman uses in his book. Suppose we are sitting near to two people playing a game of chess, but suppose we can only inspect the board and the pieces every now and then: we have no idea what the rules are, and we are trying to figure them out. After watching a few games, it will be apparent that many, many possible combinations of play are possible. Amid this plethora of possible moves, we notice that only the bishops always appear on squares that have the same color as those they started from. No matter when we look during the game, provided the bishop has not been captured, the white square bishop will still occupy a white square. So bishops "conserve" the color of the square from which they started, despite having possibly made many different moves during the game. The other pieces appear to occupy black or white squares at random as they move, but bishops stay on the same color. This is the nature of a conservation law. It is the manifestation of some underlying principle.

The conservation laws we are concerned with are a consequence of the physical laws—we do not learn anything fundamentally new, but such information is often hidden behind the detail. In the game of chess, the conservation of color by each bishop is immediately apparent from the rules: Bishops only move along diagonals; consequently, they remain on the same color. The other chess pieces do not have to move along diagonals, so they can move to squares of either color.

In physics, the principles that emerge from the equations are often quite subtle but enormously powerful. They frequently allow us to extract useful information without having to solve differential equations either explicitly or precisely for all the detail. In other words, they help to simplify problems or identify patterns even when there are many billions of variables interacting in very complicated ways.

Just as in a game of chess, where the various possible moves that the pieces are allowed to make place constraints on a player's strategy, so too we find that conservation laws in nature place constraints on how systems evolve. The existence of a conservation law means that quantities that might otherwise vary independently from one another do not do so in practice. This often helps us to spot patterns in otherwise complex behavior because even if the detailed motion of a system is ever changing, the changes have to take place in such a way that the conservation laws hold. One of the most familiar examples of a conservation law in action is that of a pirouetting ice skater. When ice skaters pirouette, they spin faster when lowering or drawing their arms toward their bodies, due to conservation of their *angular momentum*.

Angular momentum is related to the notion of momentum of an object moving in a straight line—we feel this momentum, measured as the product of the object's mass and velocity, when the moving object bumps into us. If an object is moving in a circle at a constant rotation rate, then it possesses a momentum that is proportional to the distance of the object from the axis about which it is circling. Angular momentum is defined as the product of this momentum times the radius of the circle. If we think of a disk rotating at a constant rate, then we know that a point near the edge of the disk moves through space faster than a point closer to the center, even though the points will make the same number of revolutions or complete turns in a given time. Rotating galaxies (figure 5.3) and freely rotating discs (for example, a Frisbee) share this property of conservation of angular momentum.

Returning to the ice skater, no matter how complicated the dance routine may be, if the skater spins and changes the position of their arms relative to their body, then this will affect their rotation in a predictable way. If the equations imply that the total angular momentum of a rotating system remains constant, then the distribution of matter in such a

Figure 5.3. Conservation of angular momentum helps us understand the motion of stars in a galaxy. Here millions of glowing suns wheel as if on a disc spinning around its axle. By using conservation laws we can study the rotation of galaxies without having to compute the motion of all the individual stars. © Konstantin Mironov.

rotating system and the rate of rotation of that system do not vary independently of one another.

As we mentioned in chapter 2, the early pioneers of our story used conservation laws intuitively. Halley realized that air mass had to be conserved—if air flows out of a region, it must be replaced by air flowing into that region, and as the air flows, it cannot suddenly disappear. This in itself may not seem terribly profound, but it is a principle that is quantified by one of the equations of fluid motion—the principle of *conservation of mass*. This principle, together with the notion of buoyancy, as explained by Archimedean hydrostatics, provided Halley with the basis for his theory of the general circulation of heat from the tropics to the poles. The further step taken by Hadley brought in the above angular momentum idea, in his remark to the effect that the northeasterly and southeasterly winds near the ground within the tropics and subtropics must be "compensated for elsewhere," otherwise there would be some net change—a slowing down—in the spinning motion of the Earth. This seemingly innocuous remark concerning the need for a large-scale atmospheric motion to compensate for the winds in the tropics is profound. It shows us Hadley's grasp of the problem. The lawyer had recognized an aspect of the principle of conservation of angular momentum for the rotating Earth-atmosphere system: the idea that there should be no net torque (or "twisting force") between the Earth and the atmosphere implied the necessity of the westerly winds of middle latitudes. Indeed, so powerful are the notions of conservation that neither Halley nor Hadley needed to write down a single equation, nor a solution, in either of their papers in order to explain their theories.

We next demonstrate why Kelvin's theorem is so useful. We start by extending the idea of the angular momentum of a moving zone of air circling our planet. Consider a band or belt of air moving eastward and extending all the way around the Earth at the latitude of the Tropic of Cancer. Suppose that this band of air is gently drifting northward. Then we ask the question "how might the strength of the wind change?"

Calculating the motion of such a mass of air is an interesting meteorological problem (not least because of Halley and Hadley's theories of general atmospheric flow). By using Kelvin's circulation theorem, we can deduce important general facts about the motion without having to solve the detailed forces, wind speeds, and pressures that make up

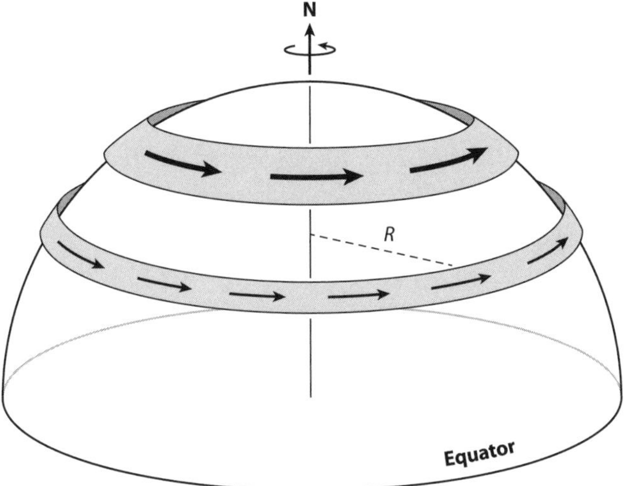

Figure 5.4. The entire belt of eastward moving air at speed V drifts northward. The belt is at distance R from the axis of spin of the Earth. Because the angular momentum per unit mass $2\pi RV$ remains constant, as the length, $2\pi R$, of the belt decreases, the speed, V, must increase, so we see stronger westerlies. If we think of the air mass in this belt, then we conserve the total angular momentum of the air mass. If instead we think of individual air parcels, then each parcel's angular momentum is also conserved. The total angular momentum can be represented in terms of circulation, as in the Kelvin and Bjerknes theorems. This often turns out to be more helpful.

the equations of motion. We first ignore the effect of temperature variations in the atmosphere. Then Kelvin's theorem can be used—at least to a good approximation. It immediately tells us that the northward drifting air belt will develop cyclonic motion (counterclockwise looking down from a satellite above the North Pole), and westerly winds will be produced. Since total circulation is conserved as the eastward-moving air drifts northward, the diameter of the band decreases and the band moves faster toward the east (see figure 5.4).

The angular momentum of a whole belt of air rotating around our planet is a useful concept when explaining the occurrence of trade winds (or variations in the jet streams, as we discuss later). But usually we wish to analyze more localized phenomena, such as cyclonic storms. Circulation or vorticity on air parcels is then much more convenient. We use the circulation theorem to show that swirling masses of air that are contracting toward their local vertical axis of rotation will acquire local cyclonic circulation (that is, counterclockwise rotation observed from a satellite overhead). The reverse rotation happens when the swirling masses of air are expanding.

When we change our viewpoint from considering the motion of the atmosphere over distances of several thousands of kilometers to more regional scales of a few hundred kilometers, we find that air mass behavior still conforms to these rules. This swirling motion is typical and observed in midlatitude high- and low-pressure systems. A number of other important rules can be deduced from these circulation theorems, and Bjerknes's extension of Kelvin's result to include independent variations in pressure and density, due to such effects as heating, has far-reaching consequences. These conclusions about circulation assume idealized flow conditions and ignore—indeed, are mostly independent of—such things as randomly gusting local winds.

One might correctly think that, given the vagaries of the weather, it is extremely unlikely that these idealized situations will occur at all. However, it turns out that the circulation theorem can be used with surprising accuracy, as we saw in chapter 1. A local version of this theorem underpins certain "rules of thumb" that forecasters use, and lies at the heart of Rossby's work. The rotation of air masses has a much more detailed or local conservation law, quietly organizing the flow—the law that Bjerknes had realized was so important and that Rossby would so triumphantly demonstrate (and to which we turn in the final section of this chapter).

The intellectual journey to Rossby's triumph involves several key stages. The first step was to understand how a large-scale weather pattern will change over a week or so, and the challenge was to devise a mathematically tractable model that captured the essence of the problem.

The Expediency Man

Rossby's quest for the breakthrough mathematics of weather began at MIT in Boston. During the 1930s Rossby and his colleagues were studying charts on which they had drawn the pressure patterns created by averaging the fluctuations in pressure over five-day periods. By studying these averages, they removed the more rapid fluctuations created by the motion of individual lows and highs on a day-to-day basis. The resulting charts revealed the major "highways," or tracks, of the lows, as is apparent in the chart in figure 5.5b. That is, if we look at the North American region and choose a period of weeks to allow the passage of several lows, the averaged pressure contours on Rossby's chart would be roughly parallel and define the tracks of these systems. Mostly, the patterns are larger than the dimensions of the individual weather systems. Lows do not all follow precisely the same tracks, and highs meander too. Rossby referred to the patterns as "centers of action" in the atmosphere.

The winter charts normally showed at least five such centers: the Icelandic and Aleutian lows, and the Azores, Asiatic, and Pacific highs. It was noticed that one or more of these centers frequently broke into two parts. But when charts were plotted of the pressure variations at higher altitudes, it was seen that the "centers" were no longer identifiable as such and appeared merely as local undulations in the prevailing pressure distribution. Eight kilometers up in the more rarefied atmosphere, most midlatitude weather appeared more like a wavy ribbon encircling our planet, as figures 5.8 and CI.6 show.

The discovery of these patterns immediately poses the following questions, summarized by Rossby in 1940 as follows: "do certain preferred patterns exist which are more readily established than others, and when will an arbitrary flow pattern tend to remain stationary and when will it change or move?" Quantitative answers to these questions would provide valuable information for forecasters who sought to predict the weather for the week ahead. Bearing in mind Richardson's failure fifteen years earlier to predict the change in pressure over a six-hour period at only two locations, we can begin to see the magnitude of the task Rossby was setting the group at MIT: they could not begin to address these questions by using Richardson's model and his numerical techniques.

The way Rossby set about calculating how the pressure patterns would change was quite amazing. Richardson had carefully set out all the equations for all the variables and included many, if not most, of the phenomena that affect them. Richardson even discussed including dust as a variable in his equations because it was known to provide points for raindrops to condense on. In total contrast, Rossby pulled out Occam's proverbial razor and went to work on the governing equations with a vengeance. Rossby wanted to create a model atmosphere that reflected the weekly averaged behavior of our actual atmosphere, a bit like the climate statistics that we can look up to see if a place

Figure 5.5a. This satellite image shows the cloud associated with a trio of low pressure systems over the north Atlantic. There is a massive cloud band associated with the warmer air, the warm and occluded fronts associated with the low with a pressure of 977mb as shown in figure 5.5b. Behind the cold front is the stippled pattern of much smaller clouds that give rise to local showers. © NEODAAS / University of Dundee.

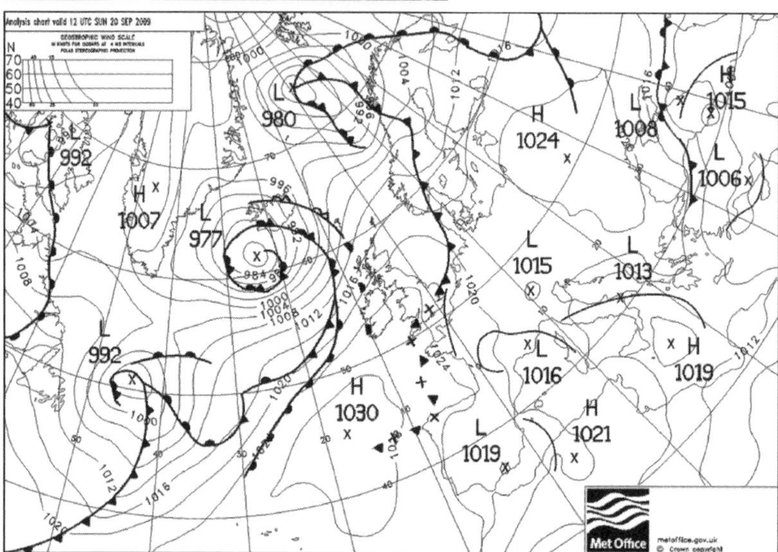

Figure 5.5b. This is the synoptic chart for the situation shown in figure 5.5a and depicts the key features of interest to a forecaster—the position of the fronts and the pressure contours (isobars). The path or track being followed by the family of three low pressure systems with central pressures 992, 977, and 980mb, is what Rossby was attempting to quantify and understand.© Crown Copyright, Met Office.

is warm and dry in autumn, or wet in summer, and so on. To focus on the mathematics of "average weather," Rossby ignored the effects of friction, the Earth's geography, and the effects of excessive or varying solar radiation and heat. Rossby also ignored rain and the effects of condensing water vapor.

Of course, by making all these simplifications, his model atmosphere might appear at first sight to be very uninteresting indeed. The daily driving force of the Sun's heat and the presence of water vapor to make clouds and rain are crucial ingredients of weather as we experience it. But Rossby's purpose was to isolate the mechanisms that create the overall pressure patterns that move weather across the great oceans and the continents; these were Rossby's "centers of action," and he wanted to get to the heart of what creates large planetary-scale motion.

As we see in figure CI.5, the troposphere (the lowest part of the atmosphere where most weather—such as cloud and rain formation—takes place) is a relatively thin layer surrounding Earth, and we really only need to consider, say, the first ten kilometer depth of atmosphere. Rossby assumed that this ten-kilometer-thick layer had constant density so that no convergence of air was allowed. In contrast, in the horizontal he considered much larger scales: a typical cyclone has a diameter of roughly eight hundred kilometers, and these sorts of distances define our notion of "large-scale" in the horizontal.

By assuming the atmosphere has a constant density, Rossby eliminated sound waves in his model. Sound waves are caused by air being compressed sufficiently quickly and, of course, we are aware of their presence all the time, from hearing birds to the roar of traffic and distant thunder. But these sound waves, although part of the complete solutions of the unapproximated governing equations, are of no immediate interest to forecasters.

Rossby did, however, retain one crucial ingredient—the ingredient that would control the "weather" in this very simplified world: the variation of the Coriolis term with latitude. As we mentioned in chapter 2, Ferrel realized that the rotation of the Earth has a major effect on the airflow in weather systems, as the Coriolis term makes winds blow more along the pressure contours than across them. So the variation of the Coriolis term would matter for such eastward moving weather systems that also drifted in a north–south direction. By retaining only the variation of the Coriolis term, Rossby had a toy model whose solutions were about to revolutionize understanding of the motion of such large-scale

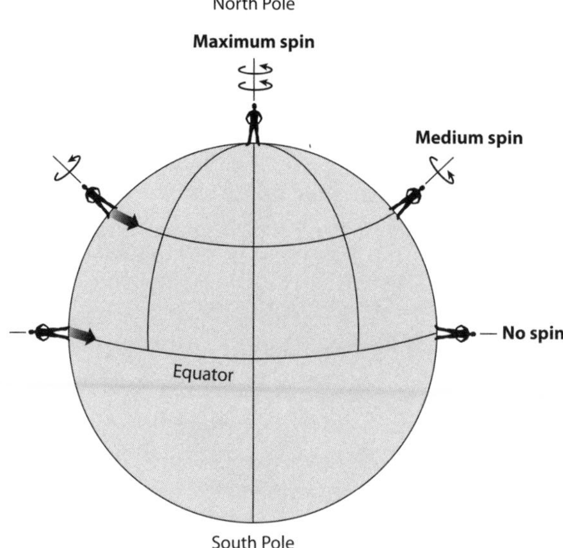

Figure 5.6. A drawing of the Earth showing the variation of planetary spin in a locally vertically upward direction. When we stand at the North Pole, we experience a maximum spin (shown by the curved arrows above our figure, who is pirouetting like a slow-motion ice skater). In contrast, when we stand at the equator, even though we are being flung around in space as the Earth rotates, we are not spinning about a local vertical axis. The Coriolis term that appears in the equations of motion (and that is retained in Rossby's model) is $2\omega\sin\varphi$, where φ denotes latitude and ω is the magnitude of the Earth's angular velocity. Thus, at the equator ($\varphi = 0°$) this term is zero while at the North Pole ($\varphi = 90°$) it is equal to 2ω. We also show a latitude, say, 45° north, with medium spin.

weather systems. He provided an explanation of how the rotation of the Earth could produce simple patterns in the weather.

The Rossby recipe for isolating the dominant process governing the large-scale motions of weather systems involved combining these simple assumptions with more than a dash of insight. Rossby focused on the vorticity, and the result was quite remarkable. By using a conservation law, he showed that the vorticity distribution is what determines the stationary or progressive characters of the large-scale motions, as we discuss in the next section.

From the outset Rossby's goal had been to explain the large-scale patterns observed in the pressure charts, but his route to finding the explanation involved removing pressure from the equations. As we saw in tech box 3.2, when we differentiate and subtract the simplified horizontal momentum equations in just the right way, we get a vorticity theorem. When we calculate the rate of change of the vorticity of the flow, the pressure gradient force disappears from the resulting equation of motion. Put very simply, the pressure gradient force pushes the fluid parcels and cannot make them spin. So these forces cannot directly change the vorticity of a fluid parcel. We now look at a consequence of this vorticity conservation law.

Rossby's Stately Waltz

Rossby used the conservation law for vorticity in an atmosphere of uniform depth and temperature to derive one of the most celebrated and useful formulas in meteorology. To see how it gave him understanding of one of the most basic features behind the creation of weather systems, we first take a closer look at vorticity itself.

Meteorologists talk of "relative vorticity," "planetary vorticity," and "total vorticity," which makes it even more confusing at first. The relative vorticity is the measure of vorticity relative to the surface of the Earth. This is similar to the spin that an ice skater has when pirouetting at one point on the ice. We ignore the fact that the ice, as part of the Earth, is actually moving too (at 50 degrees north, say, at New York or Chicago, the ice rink is traveling eastward at roughly 1,000 kph due to the rotation of the Earth). The planetary vorticity is the vorticity at sea level due to the Earth's rotation. That is, if the atmosphere were at rest relative to the Earth, so that an observer would experience total calm, then such a state of the atmosphere would still have nonzero vorticity (because it is still spinning around with the rotation of the Earth, as a spectator in a spacecraft, or on the Moon, would see); so we call this planetary vorticity.

Planetary vorticity is solely determined by latitude: it is zero at the equator and maximum at the poles, as shown in figure 5.6. Meteorologists and oceanographers need the notion of total vorticity—a measure of the total rotation, or spin, of a fluid parcel—because this is what the conservation law refers to. Total vorticity is the sum of the vorticity relative to the rotating Earth and the vorticity due to the Earth's own rotation about its axis (the planetary vorticity). It is difficult to measure, compute, and predict the relative vorticity, so our aim is to use a

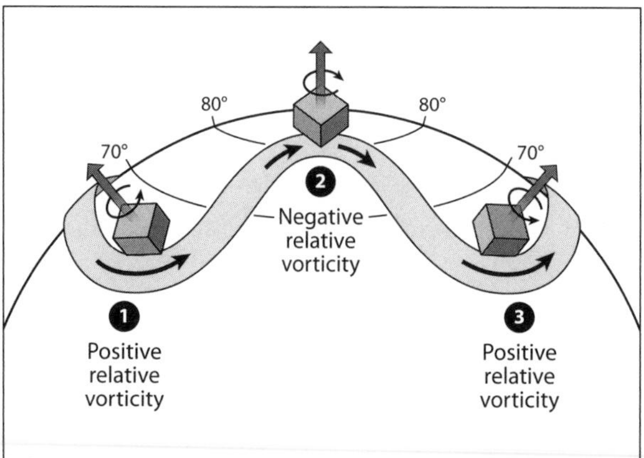

Figure 5.7. Consider an air mass that has no relative spin at, say, 70° north. Now it moves 10° southward, keeping the total spin at the same value. The planetary spin at (1) (about 60° north) is smaller than at latitude 70° north as (1) is nearer the equator; hence, the relative spin is positive at (1), so that the total spin remains constant. The extra local spin then moves the air mass at (1) back toward the pole. Similarly, the planetary spin at (2) (about 80° north) is greater than at latitude 70° north as (2) is nearer the pole; hence, the relative spin is negative at (2). This negative local spin then moves the air parcel back toward the equator. The whole cycle repeats, and in principle the air mass could continue forever wobbling between 60° and 80° north as it travels eastward.

theorem, a conservation law, for the total vorticity and hence find the relative vorticity of an air parcel as the difference between two things that are both more readily measured and known—the total and the planetary vorticities of that air parcel.

In a horizontal layer of idealized atmosphere at a constant temperature, the total vorticity of a horizontally moving parcel of air obeys a conservation law. Therefore, its relative vorticity must vary in a definite fashion in order to compensate for the variation in the planetary vorticity of the air parcel: the sum of the vorticities must remain constant. This imposes certain constraints on the tracks or trajectories that may be followed by any column of air, and certain preferred flow patterns will then be established. That is, winds carrying air parcels to different latitudes cause the relative spin or vorticity to change in a way that compensates for the known change in planetary vorticity. This change then feeds back on the winds, causing the air parcels to retrace their paths across the lines of latitude. In a perfect world, the air parcels would oscillate back and forth forever, as indicated in figure 5.7, just like a swinging

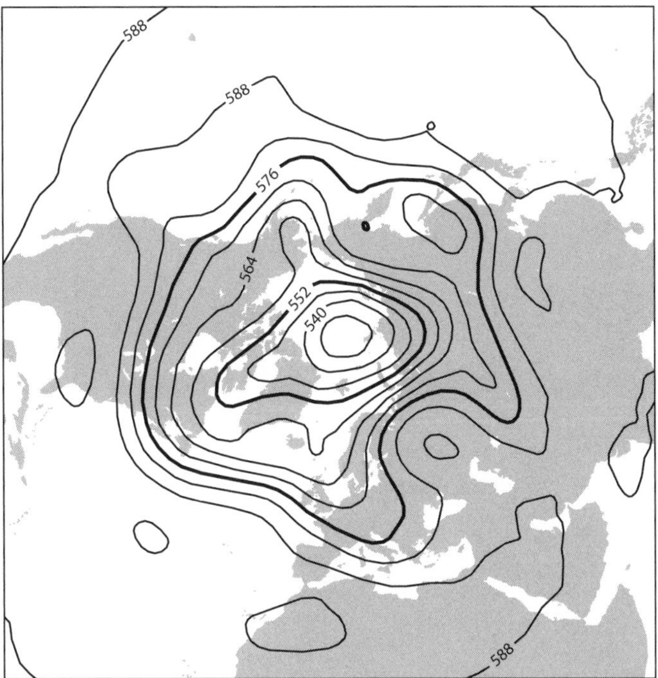

Figure 5.8. A top-down view, from above the North Pole, of Rossby waves. The contours show the elevation, in decameters, of the 500mb pressure surface. This is usually referred to as a height field: the lower the surface, the colder and denser the air. This is an ECMWF height field analysis averaged over 10 days July 21–31, 2010. This was the period of the extreme Russian heat wave (see ridge over central Russia) and the major flooding in Pakistan. © ECMWF. Reprinted with permission.

pendulum. Rossby calculated a simple formula for this oscillation, as described in tech box 5.1. The foregoing reasoning shows us once again how conservation principles can help us solve otherwise complicated nonlinear problems. It explains certain elements of the northern hemisphere airflow depicted in figure 5.8.

Rossby solved his equations, and the mathematical patterns revealed—for the first time in history—useful approximations to the pressure patterns first discovered by Solberg in his Bergen model of the polar front. The waves were very large scale, extending thousands of kilometers from west to east. These mostly westward propagating waves in the vorticity, associated with eastward-moving weather, became known as *Rossby waves.* Rossby was able to provide forecasters with a very simple equation

Tech Box 5.1. Rossby Waves in the Upper Atmosphere

The equation that underpinned Rossby's model was the conservation of total vorticity, ς, calculated on an air parcel:

$$D\varsigma/Dt = D\varsigma_s/Dt + \beta v = 0,$$

where ς_s denotes relative vorticity, β is the rate of change of the Coriolis term with latitude ($\beta a = 2\omega\cos\varphi$, where a denotes the radius of the Earth), and v is the north wind, with the notation used in chapter 2. If there were no rotation of the Earth ($\omega = 0$), then this equation is the same as Helmholtz's theorem, which is Kelvin's theorem on small air parcels. The equation as it stands is nonlinear because the total derivative depends on following the air parcel, which requires knowing the wind velocity. Rossby linearized this equation about a zonal flow of uniform eastward speed, U, and considered small perturbations u' and v' about this zonal flow. In the case of a simple sinusoidal perturbation of wavelength, L, which is independent of latitude, the perturbation is given by an expression of the form

$$v' = \sin[(2\pi/L)(x-ct)]$$

where c denotes the phase speed. This phase speed is the speed of propagation of the wavy disturbance to the east. Substitution of this perturbation into the linearized equations shows that

$$c = U - (\beta L^2)/(4\pi^2).$$

This is Rossby's famous equation. The waves become stationary when $c = 0$, that is, when $U = (\beta L^2_0)/(4\pi^2)$, or when

$$L_0 = 2\pi\sqrt{(U/\beta)}.$$

Therefore waves of length greater than L_0 travel westward, and shorter waves travel eastward—a very simple criterion but one that proved invaluable to forecasters.

The total number, n, of waves around the circle of latitude, φ, is given by

$$nL = 2\pi a \cos\varphi.$$

Rossby's predictions are in good agreement with observations, as indicated by figures 5.8 and 5.9.

for computing the speed of such waves—probably the most celebrated equation in twentieth-century meteorological literature.

The wavelength of a Rossby wave, or one "undulation," generally has an east–west extent of about five thousand kilometers (roughly across the northern Atlantic Ocean—see figure 5.8). By studying whether the spacing of the troughs and ridges is greater than or less than that wavelength calculated for the average winds (Rossby's calculations are shown in figure 5.9), it is possible to estimate their subsequent motion. Rossby was able to furnish the meteorologist with a simple formula for the speed of propagation of these waves in terms of the wavelength so

TABLE II

STATIONARY WAVE LENGTH IN KM AS FUNCTION OF ZONAL VELOCITY (U) AND LATITUDE (φ)

φ \ U	*4 m/sec*	*8 m/sec*	*12 m/sec*	*16 m/sec*	*20 m/sec*
30°	2822 km	3990 km	4888 km	5644 km	6310 km
45°	3120	4412	5405	6241	6978
60°	3713	5252	6432	7428	8304

TABLE III

VELOCITY DEFICIT ($U-c$) AS FUNCTION OF NUMBER OF PERTURBATIONS (n) AND LATITUDE (φ)

φ \ n	*2*	*3*	*4*	*5*	*6*	*7*
30°	150.7 m/sec	67.0	37.7	24.1	16.7	12.8
45°	82.0	36.5	20.5	13.1	9.1	6.7
60°	29.0	12.9	7.3	4.6	3.2	2.4

Figure 5.9. Tables II and III from the Rossby et al. paper of 1939, showing speed/wavelength estimates. With reference to figure 5.8 Rossby could successfully predict that a wave number 5 pattern at about 45° north has a velocity deficit of about thirteen meters per second, which implies a particular westward motion. © *Journal of Marine Research*. Reprinted with permission.

forecasters could estimate where and at what speed the cyclonic weather system would move next.

It turns out that the meanderings of a large river of air, often flowing eastward at more than 300 kph and about 5–8 km up in the atmosphere, determines significant weather on our planet. Such a river is known as a *jet stream.* Jet streams are fast-flowing, relatively narrow air currents located high in the troposphere, as revealed by the "river of cloud" in figure 5.10. The major jet streams in the Earth's atmosphere flow toward the east. Their paths typically have a wavy shape and may split into two or more parts. Typically, the northern hemisphere and the southern hemisphere each have polar and subtropical jets. The northern hemisphere polar jet is situated over the middle to northern latitudes of North America, Europe, and Asia, while the southern hemisphere polar jet mostly circles Antarctica all year round.

Figure 5.10. Jet stream, Maritime Provinces, Canada, May 1991. The northern hemisphere jet stream can be seen crossing Cape Breton Island. During the winter months, the path taken by the jet stream over the United States and southern Canada can have a significant impact on the weather conditions. Courtesy of NASA.

The 576 and 552 contours encircling planet Earth in figure 5.8 show typical jet streams. The response of the jet streams to changes in the state of our atmosphere could significantly influence weather patterns in the future.

Of primary importance is the interplay between the Rossby waves, the polar front, and the jet streams—an interplay that is crucial for weather forecasting. A jet stream moving across North America, together with associated weather systems, is apparent in a conventional weather map, such as figure CI.6. The similarities between figures 5.7 and 5.8 were what Rossby focused on (although jet streams were not discovered until the Second World War)—the common features showed the way that rotation constrained actual time-varying, complicated, fluid motions.

Rossby's key ideas and results were presented in two papers in 1936 and 1939, and he published a paper in 1940 summarizing the findings. In the introduction to the 1940 paper, he remarks: "Most of the results presented below are readily obtained with the aid of Bjerknes's circulation theorem, which has been available to meteorologists for the last forty years. Under these circumstances it is rather startling that no systematic attempts have been made to study the planetary flow patterns in the atmosphere." His expediency resulted in a model that has become the starting point of theoretical meteorology. Extensions to Rossby's work underpinned forecasting practice for many decades.

Although Rossby ignored a great deal (indeed, nearly all) of the physical detail in deriving his equation, he kept conservation of mass and conservation of vorticity on each fluid parcel; his idealized models also conserved total energy. Thus, Rossby identified the key "backbone" of weather. As a person's backbone needs to support the body weight against gravity, and muscles act to move the body around this state of basic balance, so these conservation principles provide the basic state of balance in the atmosphere and control relative motions. Once we are "balanced," the resulting theory begins to provide an understanding of the ever-changing weather.

But, as Bjerknes realized in 1898, many flows of interest, such as the fishing waters off the Swedish coast and the layers of our atmosphere where clouds form, have both temperature and density variations. We need a more elaborate version of both the vorticity conservation idea and the highly idealized Rossby wave theory. This more realistic theory is known as conservation of *potential vorticity*, and it too had its origins in Rossby's work of the 1930s.

The Invisible Choreographer

Rossby's quest for a succinct mathematical description of large-scale weather patterns led him to define a concept that is arguably one of the most powerful in modern meteorology. This concept is a logical consequence of Bjerknes's circulation theorem, and it can be deduced from the governing equations (after several lines of exhilarating calculations). The concept is called potential vorticity (or, adopting the meteorological acronym, PV). The word "potential" is used because, unlike ordinary vorticity, potential vorticity incorporates our knowledge of thermal processes at work in the atmosphere. The potential vorticity of an air parcel is calculated from the rotation of that parcel scaled by the gradient in temperature along the axis of rotation. This might seem a little complicated—and it is far from straightforward—but the payoff for studying this idea is that PV orchestrates the behavior of entire weather systems, as will be seen in figure 5.13 at the end of this chapter.

In his 1940 paper, Rossby published a first extension of his original model to allow for the horizontal convergence of airflow and the associated changes in density. At essentially the same time, Hans Ertel, a professor of geophysics in Berlin, was working on the problem of incorporating variations in temperature with the vorticity equation. The confluence of their work produced what we now recognize as PV.

The extension of PV to situations where variation of temperature is significant is facilitated by the introduction of another quantity, called potential temperature (see tech box 5.2), which we label θ (theta) to distinguish from ordinary temperature.

When we go beyond Rossby's simple model and include the heat and moisture processes that are crucial in creating rotating air patterns in the atmosphere, it transpires that it is not vorticity that is conserved but a combination of vorticity and (the gradient of) a thermodynamic variable such as potential temperature. Essentially, Rossby turned Bjerknes's circulation theorem into a statement about vorticity behavior by focusing on a flattened disc or columnar parcel of thermally isolated air lying

Tech Box 5.2. Potential Temperature Measures Heat Energy

The formal definition of potential temperature is as follows.

The *potential temperature* of a parcel of air at pressure p is the temperature that the parcel would have if it were brought adiabatically (with no transfer of heat) to a standard reference pressure, p_0, say, the standardized pressure at sea level. We use θ to denote potential temperature, and for air we have the formula

$$\theta = T(p_0/p)^{\gamma},$$

where T is the current absolute parcel temperature (measured on the Kelvin scale) of the parcel, and γ is a constant. As a balloon, or parcel, of isolated air is expanded, the value of θ remains constant only when there is no net exchange of heat energy across the surface of the balloon. So constancy of θ goes with conservation of thermal energy on air parcels as they move and change shape in a pressure-varying environment.

between layers in the atmosphere marked out by values of the potential temperature.

Potential temperature allows us to combine variations in temperature with Rossby's generalization to include converging airflows. We explain PV by considering the spin of a column of air as it shrinks in cross section (converges) and stretches in length vertically upward—see figure 5.11.

We think of these columns as a physical realization of a marked set of air parcels where rotation is a key feature of the air motion. As with our earlier discussions of the dynamics of an ice skater in terms of angular momentum, these conceptual columns play a very similar role in realizing the dynamics of the atmosphere in terms of PV. First we think of two conceptual surfaces lying nearly horizontally, one above the other, rather like blankets in the atmosphere. The lower surface is marked by all of its air having the value θ_{BOTTOM} for its potential temperature, while the upper surface has air parcels with the potential temperature θ_{TOP}.

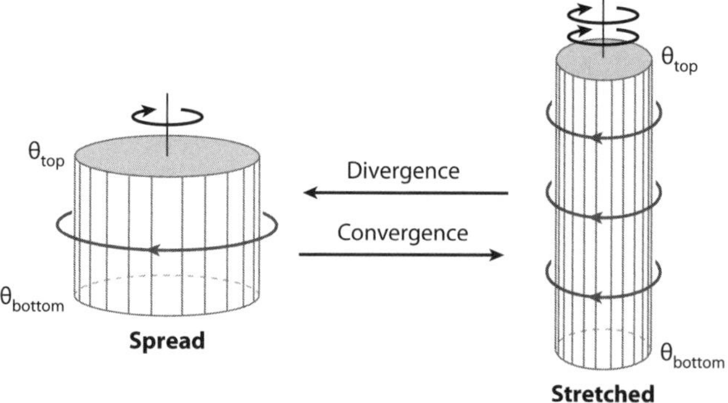

Figure 5.11. The convergence and stretching of a column of air, accompanied by an increase of its spin, is reminiscent of the pirouetting ice skater. Each column extends upward from the air surface marked out by the potential temperature θ_{BOTTOM} to that marked out by θ_{TOP}. These columns always consist of the same air parcels during the motion.

We now think of a column (of rotating air) that always extends between two θ surfaces while it moves around. When the column of air shrinks in cross section (convergence), then it must increase its vertical extent—the "stretching" in figure 5.11—since the column contains the same amount of air. Now, relative to its environment, the rate of rotation of this column—its relative vorticity—should also increase when we think of the spinning ice skater. We may guess that there exists a quantity relating the spin and convergence, which remains constant. If we recall the ice skater and remember vorticity is the spin here, then our guess for this quantity might be the ratio of the total vorticity of the column to the "height" of that rotating column. This guess turns out to be correct.

The total vorticity of the column is the sum of the planetary vorticity ς_p and the relative vorticity ς_s. (The Greek letter ζ [zeta] is commonly used to denote vorticity.) The PV of the column is then the total vorticity divided by the height of the column, H, which reads symbolically as

$$PV = (\varsigma_p + \varsigma_s)/H.$$

For atmospheric flows in which PV is constant, when H decreases (say is halved), then the total vorticity, $\varsigma_p + \varsigma_s$, must decrease in proportion (that is, is also halved). Rossby and Ertel found the conditions under which PV is constant following columns of air as they flow between the relevant constant θ surfaces: that is, the conditions when PV is a conserved quantity.

We now use this definition of PV to explain how flow over a mountain range can produce a Rossby wave pattern. As shown in figure 5.12, we consider an eastward flow blowing over a mountain range that is aligned north and south (such as the Andes). This flow is assumed to have zero relative vorticity to the west of the mountains, and we shall assume the eastward component of the flow remains uniform across the entire range. Such a flow has nonzero planetary vorticity, ς_p (because we are not at a point on the equator). As the column is blown over the range, it flattens and H decreases, so if PV is conserved, the total vorticity must also decrease. Initially, the planetary vorticity, ς_p, remains constant, so a positive relative vorticity, ς_s, is induced. As the eastward component of the flow is assumed to remain uniform, this means a northward component of the flow must be generated. Hence, our column moves

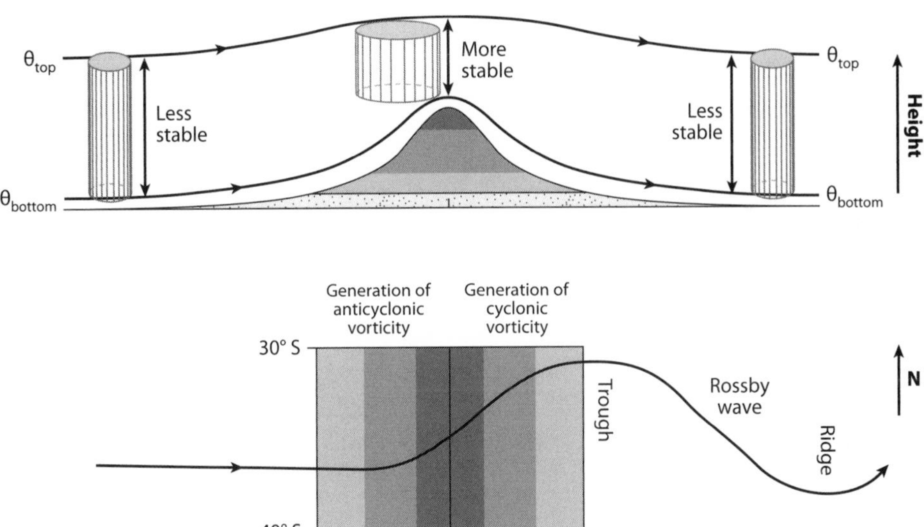

Figure 5.12. Here the effect on air parcels of crossing the Andes is to start a gentle north–south oscillation of the air as it travels eastward, and this oscillation generates a Rossby wave. Such convergence of an air column lying between two blankets of air marked by their θ values takes place when air flows from the plains over a mountain range. The reverse happens on the downwind side of the range. The change in the vorticity of the column that accompanies this often leads to characteristic cloud and weather patterns being created on the lee side of the mountains, such as over the high plains when air blows from west to east across the Andes. Adapted from a figure in B. Geerts and E. Linacre, "Potential Vorticity and Isentropic Charts," http://www-das.uwyo.edu/~geerts/cwx/notes/chap12/pot_vort.html.

north. But now the planetary vorticity, ς_p, begins to increase. As the flow reaches the eastern side of the range, the height, *H*, begins to increase again; therefore the total vorticity must decrease to compensate for this. The planetary vorticity, ς_p, has increased, so the relative vorticity must decrease in order for the sum $\varsigma_p + \varsigma_s$ to compensate for the increase in *H*. The total vorticity is decreased by creating a southward component of the flow. We have started a Rossby oscillation about some intermediate latitude, which in theory, when nothing else happens, could carry on forever as the air parcel moves eastward around the planet.

Once again we appreciate the power of conservation laws in defining types of weather systems—despite the complexity of the detailed winds, the pressure and temperature fluctuations, and the odd spot of rainfall. The statement that the total vorticity of a thermally isolated column of fluid must vary in accordance with the appropriately measured stretching and shrinking of the column is always true. The possibly quite complicated airflow is constrained by this law. Just as the pirouetting ice skater uses conservation of angular momentum to generate intricate routines by moving her arms, our column of air, although moving in intricate ways, retains the value of PV it had when we began to identify and follow it. The change in relative or local vorticity interacts with the local environment to help move our column, which then further changes the local vorticity, and so on. The conservation of PV is what allows us to cut the Gordian knot of this nonlinear feedback.

Bjerknes had shown that circulation could be generated when the temperature changed, and this is important during the development of cyclones; Rossby showed that while these changes in circulation were going on, the air was still subject to an important constraint, namely the conservation of PV. Most of the time both the vorticity variable (which depends on the air motion) and the temperature variable will be changing. Therefore, this PV constraint, that all these variables have to vary in precisely the right way to keep the potential vorticity constant following the motion, is an important controlling mechanism at work in the atmosphere—rather like a referee trying to keep a balance between the competing forces on air parcels.

The use of PV has followed two directions since the early 1940s. The first involved the further mathematical development of Rossby's theory, which is known as quasi-geostrophic theory (and which we discuss much further in the following chapter), and the second was in the analysis of weather charts themselves. Following the development over the past thirty years of numerical weather prediction based on computer solution of the full equations of motion, the use of potential vorticity for analyzing weather charts decreased. Its incorporation into computer simulations remains problematic to this day, and we shall say more about this in chapters 7 and 8.

In the 1940s and '50s, a great pioneer of the use of potential vorticity as a tool for meteorologists was Ernst Kleinschmidt. He identified the possibility of deducing the wind, pressure, and temperature fields from the PV itself. Kleinschmidt was building on the work of Ertel; in fact, both were invited to the research institute in Stockholm that Rossby founded in the 1950s. The conservative properties of potential vorticity were retained by Kleinschmidt, but the equations become more complicated. We say more about the math of PV and potential temperature in chapter 7: taken together, they provide meteorologists with a powerful diagnostic of motion in the atmosphere.

This development of "PV thinking," as meteorologists call it, brought together many strands of theoretical meteorology and practical forecasting. To arrive at the PV equation, we have to eliminate pressure and density from the equations—even the wind field is relegated to a phenomenon of secondary interest. There are also new challenges in what has become known as "PV inversion" when the wind fields and the temperature, together with the pressure, are recovered from these PV fields.

But what are the implications of these conservation principles—those for mass, energy, moisture, and PV—for numerical weather forecasting? After all, we have powerful supercomputers to crunch through the calculations that are necessary to produce a forecast from the basic equations of motion without the need to check whether the conservation laws are represented. Are these topics of interest solely to the specialist theoretician, or are they of practical value?

The answer lies in the observation that the computer models, although based on the laws of physics, do not necessarily inherit the exact consequences of these laws via the computer programs that encode them. That is, a numerical representation of the laws of motion and heat and moisture that we described in chapter 2 will not automatically or exactly represent Bjerknes's circulation theorem because of the ever-present

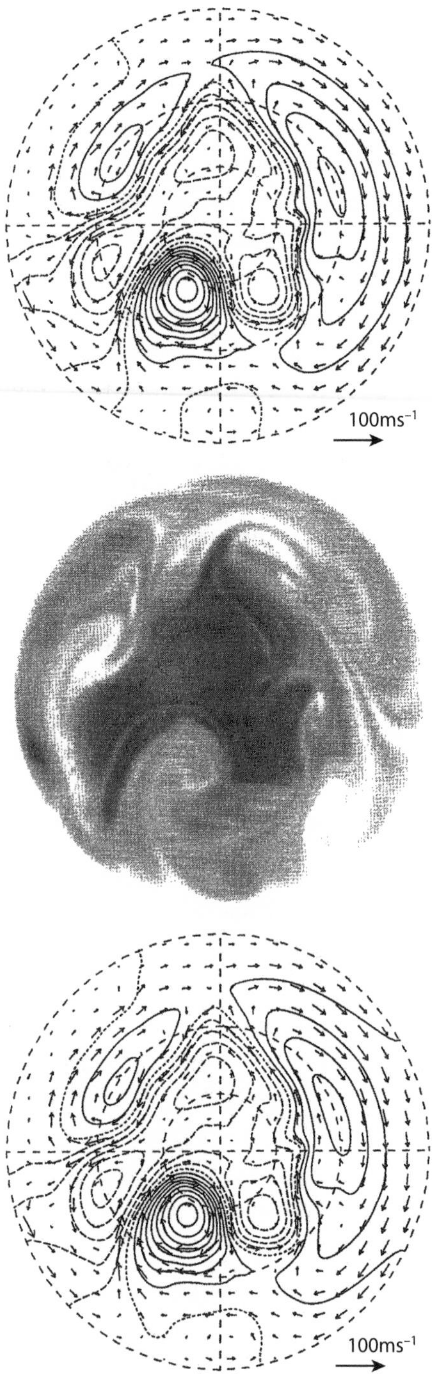

Figure 5.13. The first panel shows the wind vectors and height contours (joining points with the same value of the variable *H*) of a simulation of an idealized flow in a hemisphere. The second panel shows the PV calculated from the total vorticity and the height. The third panel shows the wind vectors and the height contours, but these have been calculated from the PV, using a method known as "PV inversion." It is difficult to spot differences between the first and third panels, and thus reinforces the view that PV encapsulates all we need to know about large-scale flow patterns. Figure reproduced from "Potential Vorticity Inversion on a Hemisphere" by Michael E. McIntyre and Warwick A. Norton, *Journal of the Atmospheric Sciences* 57: 1214–35. © American Meteorological Society. Reprinted with permission.

very small errors and approximations; consequently, the simulations do not automatically "know" about the constraints that these conservation theorems place on the development of weather systems. We have to work very hard to build such knowledge into the computer models, and this endeavor is certainly an important topic of the latest research.

In developing numerical schemes that better represent these conserved properties, we significantly reduce the way in which errors build up from irrelevance to actually making a key difference in the simulations. By identifying the important weather properties—and PV is one—we think about how to make computer calculations follow and predict the actual weather.

SIX

The Metamorphosis of Meteorology

Rossby had loosened the Gordian knot of nonlinear feedback in the motion of the atmosphere, albeit only for the large-scale jet streams in midlatitudes. Next we describe how work on both sides of the Atlantic began to change ideas about the origins of cyclones with their attendant warm and cold fronts—the features that preoccupy forecasters in many parts of the world.

Rossby's explanation of the meanderings of the jet stream was significant not least because the polar front—the battleground between warm tropical air and cooler polar air—had been considered an essential part of the Bergen School's model of a cyclone's life history for more than twenty years. This view changed dramatically in the 1940s when, using rigorous mathematics, it was demonstrated that realistic models of cyclones could be formulated without the need for a sharp temperature contrast such as that exhibited along the polar front. Between 1945 and 1955 meteorology was transformed in two significant ways: a combination of judicious physical insight with novel mathematics led to a fundamental change in understanding the formation of cyclones and fronts, while the emerging technology of electronic computing gave meteorologists and forecasters a powerful new tool for simulating and predicting weather and climate. This is the story of the genesis of modern meteorology and weather forecasting.

Brainstorming in Princeton

On April 9, 1951, Vilhelm Bjerknes died at the grand old age of eighty-nine in his hometown, Oslo, at the end of what is surely one of the most important eras for meteorology and weather forecasting. His obituary in the London *Times* read as follows:

> Professor Vilhelm Friman Koren Bjerknes, doyen of Norwegian scientists, died in Oslo on Monday night at the age of 89. Born in 1862, he was educated in Norway and Germany and was appointed Professor of Mechanics and Mathematical Physics in the University of Stockholm in 1893. He remained there until he moved to Oslo University in 1907 to fill a similar position there, having meanwhile been elected a research associate by the Carnegie Institute, Washington, a position he was to hold until 1946. After a brief period from 1913 to 1917 as Professor of Geophysics at Leipzig University, he returned to Norway to take charge of the Bergen Geophysical Institute and in 1918 founded the Bergen weather service. He held the Chair of Physics in Oslo University from 1926 until 1932, but his alert mind was not dimmed by age and he went on working long after his retirement.

His Bergen School had produced a new generation of leading meteorologists; their ideas had transformed meteorological thinking. Bjerknes would have been justifiably proud of their achievements.

Bjerknes had lived just long enough to witness the dawn of the next important era for weather forecasting—the transformation to the modern age with the beginning of electronic computing. Pencils, paper, books of mathematical tables, and the ravages of the First World War were the setting for Lewis Fry Richardson's attempt at realizing the Bjerknes vision: within a year of the end of the Second World War, the second serious attempt to use the laws of physics to produce a prediction was under way. In place of a solitary figure working in a miserable rest billet behind the trenches while waiting to help wounded soldiers, a team of leading scientists was engaging with the new technology of electronic computing in the research powerhouses of modern America.

The Second World War may have been an unwelcome diversion to scientists and their agendas, but it accelerated progress in diverse areas, and many new ideas emerged. Numerous technological advances were made, including the development of radar and computing. Several years after the war, radar began to play a vital role in weather forecasting. The development of computing machines was needed to assist with urgent problems, such as the quest to break the Enigma code. And weather forecasting in its own right had been crucial to many of the operations carried out during the conflict. The impact of weather, and the prognoses of the forecasters, on events during the Second World War is well documented: the collapse of the German eastern front in Russia in the severe winter of 1941–42; General Eisenhower's choice of the day to launch Operation Overlord based on accurate weather forecasts; and the extensive use of ships and aircraft in the western Pacific.

Lessons learned earlier during the First World War were reinforced: weather forecasting was vital to military operations and, as a result, many postwar weather research programs were initiated and even maintained by military resources. The war had led to the creation of networks of weather observing stations, which were needed to facilitate airborne operations. These networks were expanded after the war to meet the needs of civil aviation, which grew dramatically. Meteorologists now had widespread data on the state of the atmosphere, leading to a more worldwide view of weather. Even theoretical meteorology was given a significant push when, following negative comments from the military, a new department at the University of Puerto Rico was opened by Rossby in 1943 to understand tropical weather and to train young people to forecast it better.

In earlier chapters we mentioned how understanding the trade winds and ocean currents had transformed journeys by sea in the nineteenth century, and we mentioned how the beginnings of air travel in the 1920s and 1930s drove research forward to understand airflows in the middle atmosphere. Once again a revolution in mass passenger transit—this time to jet-powered aircraft flying higher than airships and propeller-driven planes, and regularly crossing the large distances of the Atlantic, Pacific, and Indian Oceans—would drive a need for better prediction and worldwide coverage, both at ground level and in the upper reaches of the atmosphere. In the Second World War, planes flying west across the Northern Pacific had sometimes been caught in very adverse winds, and they would eventually turn back when fuel shortage meant that they would be unable to reach their destinations. Surfing the jet stream can now take a significant amount of time off a long-distance flight, and modern airlines need good predictions of jet stream positions to take advantage when planning their routes.

Our story behind the modern use of computers in weather forecasting begins at the end of August 1946, at the Institute for Advanced Study in Princeton, New Jersey. It was here that one of the most important conferences in the history of weather forecasting took place, and it was instigated by one man. Earlier, on May 8, John von Neumann, professor of mathematics at the Institute for Advanced Study, wrote a proposal to the Office of Research and Inventions (Navy Department) seeking support for a project whose aims were "an investigation of the theory of dynamic meteorology in order to make it accessible to high speed, electronic, digital, automatic computing." The writing of the letter was in itself remarkable for several reasons. First, von Neumann was an outstanding pure mathematician—a species popularly believed to eschew practical problems—and therefore one of the last people we would consider likely to become involved in such an enterprise. Second, if the Navy Department were keen to exploit the potential of modern computing power, why choose meteorology and weather prediction as the definitive, benchmarking problem?

The first conundrum may be answered by remarking that von Neumann was also an outstanding logician, applied mathematician, and engineer who had been involved with early work on stored programming. This technique held the key to writing computer programs in such a way that the same program could be applied repeatedly to problems in which the input data might vary from one specific task to the next—such as weather forecasting. The second question has a less clear-cut answer, but von Neumann—a veteran of the Manhattan Project, which developed the first atomic bomb—was convinced that some of the most intractable mathematical problems lay in fluid mechanics. He wrote: "Our present analytical methods seem unsuitable for the important problems arising in connection with the solution of nonlinear partial differential equations and, in fact, with all types of nonlinear problems of pure mathematics. The truth of this statement is particularly striking in the field of fluid dynamics. Only the most elementary problems have been solved analytically in this field."

Von Neumann was not an archetypal professor of mathematics: he had originally trained as a chemical engineer and, while teaching in Germany in the 1920s, the cabaret-era Berlin nightlife held a special appeal for him. In America, his eminent position at the Institute for Advanced Study, his love of parties, and the high life of the leading social circles meant that he soon became a doyen of society. Von Neumann wanted a real scientific challenge for the modern computer, and weather forecasting would be it. He first had to convince government and other corporate bodies of the viability of a project for which he

Figure 6.1. John von Neumann (1903–57) was born in Budapest, went to Princeton at the age of twenty-seven, and at thirty became one of the six original professors, along with Albert Einstein, at the newly founded Institute for Advanced Study. He held this position for the remainder of his life. Von Neumann's rapid rise to such a prestigious position reflected his genius. By the time he moved to the United States, he had already single-handedly formulated the rigorous mathematical foundations of the newly discovered theory of quantum mechanics—the physics of atomic and subatomic matter. Courtesy of U.S. Department of Energy.

sought considerable funding. So von Neumann combined genius with pragmatism, and above all he was eager to take on the intellectual, organizational, and financial hurdles that would beset the new venture in meteorology.

On January 11, 1946, an article appeared in the *New York Times* announcing that "plans have been presented to the Weather Bureau, the Navy and the Army, for the development of a new electronic calculator, reported to have astounding potentialities, which, in time, might have a revolutionary effect in solving the mysteries of long-range weather forecasting." Von Neumann's decision to focus on meteorology was influenced by his understanding of the mathematics of the problem together with an appreciation of its importance to the military. According to Philip Thompson, one of the pioneers of numerical weather prediction in the 1950s, von Neumann "regarded [the weather forecasting problem] as the most complex, interactive, and highly nonlinear problem that had ever been conceived of—one that would challenge the capabilities of the fastest computing devices for many years." As a hawk, von Neumann wanted his country to be at the cutting edge, and weather forecasting needed improving.

Rossby became involved after he visited von Neumann in Princeton to talk about his plans, and subsequently wrote to the director of the national weather bureau recommending that a project be started, especially in view of "Professor von Neumann's outstanding talent . . .

it would be desirable for us to encourage his continued interest in meteorology." Von Neumann had urged that, although the institute's own computer would not be ready for several years, the meteorology project should go ahead as soon as possible because the theorists still needed to evaluate the basic meteorological problem—the failure of Richardson's forecast had not been forgotten—before attempting to actually program a computer. The correct program design would be crucial, and von Neumann knew of a number of important developments in numerical analysis that were vital to the success of the computing project. Even with a partial understanding of the cause of Richardson's failure, the basic equations were still too formidable for the early electronic computers: someone had to come up with an alternative mathematical formulation of a weather prediction problem that the new machines could handle.

The meeting held on August 29–30, 1946, was entitled simply "Conference in Meteorology," but it was really the first conference on the use of computers for numerical weather prediction. George Platzman, in his historical review, commented that it is not uncommon for people participating in or witnessing an event that later becomes recognized as a great moment in history to be quite unaware of its importance on the future course of human affairs. However, for the twenty or so elite of the meteorological community gathered in Princeton under von Neumann's organization, the significance of the proposal for a coordinated program to pursue the goal of predicting the weather by automatic computation was very clear. Rossby had played a crucial role in advancing the mathematical theory of weather patterns over the previous decade, his Chicago School was an international success, and it was therefore natural to assume he would play a leading role in the conference. But Rossby was already in the process of relocating to his homeland to establish a center for a new research team back in Stockholm. Notwithstanding this, Rossby was absolutely clear about the importance of the Princeton project, and he followed the ensuing developments with keen interest.

Also present at the meeting was a brilliant young meteorologist, Jule Charney. Charney was just out of graduate school, having completed his thesis on a new mathematical approach to understanding the origins of cyclones. His doctoral paper, when published in 1947 in the American Meteorological Society's *Journal of Meteorology*, took up almost the

Figure 6.2. Jule Gregory Charney (1917–81) was born on New Year's Day 1917 in San Francisco. Writers, artists, and other scholars were represented in the families of both his parents, and therefore intellectual values were very much a part of Charney's upbringing. His mother was a talented pianist. Although music was to become the love of Charney's life, he never trained as a musician, in contrast to one of his childhood playmates, Yehudi Menuhin. Among Charney's many honors was the Clarence Meisinger award for the outstanding young meteorologist in 1949. Courtesy of the MIT Museum.

entire issue, and its publication heralded the arrival of a leading figure in the science of weather forecasting. (This journal had been founded by Rossby; in fact, Rossby had helped start the University of California at Los Angeles (UCLA) training program for wartime meteorologists under the guidance of Jack Bjerknes and Jorgen Holmboe, the latter supervising Charney's doctoral work.) But in the summer of 1946, Charney was essentially unknown. It would be his involvement in von Neumann's project that would propel him to the forefront of his subject; in particular, his role in finding a way to convince everyone, not least themselves, that they could avoid the pitfalls of Richardson's calculation was crucial.

At this 1946 Conference on Meteorology, no great telling remarks were made or answers given. But the issues were aired, people got to know one another and to listen to different views, and von Neumann began to contemplate the essence of a successful "mountain climbing team": could they ascend the "Everest" of the first computer weather forecast?

Rossby, using the conservation of total vorticity, took the initial steps along the road to explaining how and why the large-scale midlatitude atmospheric flows give rise to the relatively slowly evolving high- and low-pressure weather patterns that we observe. In his doctoral thesis, Charney worked out how individual weather systems start to develop on the large-scale eastward blowing zonal wind patterns of the upper atmosphere that Rossby had studied. In August 1946 the answer to the

Figure 6.3. Charney's fellow UCLA students enjoyed a cartoon in the student newspaper at his expense. The caption of the original shows what the intense doctoral student is saying to the lady in the evening gown: " . . . and since these are hypergeometric differential equations with logarithmic singularities . . ." Fortunately, Elinor was sympathetic, and they were married in 1946. *The Atmosphere: A Challenge.* © American Meteorological Society. Reprinted with permission.

sixty-four thousand dollar question looming over those at the conference lay just one step further in the methods Rossby and Charney had developed, yet neither of them realized it at that time. The issue was not how to follow in Richardson's footsteps, confronting all the details and difficulties, but how to find a completely different route to their goal. It was not so much a question of *how* to predict; rather, it became a question of *what* to predict—and math would play a crucial role in coming up with an answer.

There is a clue in Rossby's discovery of the large-scale meandering waves that carry his name. It turns out that wave motion is ubiquitous in the atmosphere. This may not be entirely obvious at first. Although we are well aware of sound waves, there are many other types of wave motion, each with its characteristic spatial scale and frequency. As shown in figure 5.8, Rossby waves have spatial scales of thousands of kilometers, while figure 6.4 shows much smaller-scale waves in altocumulus cloud, which may extend over distances of tens of kilometers. The imperative is to identify which waves are important for forecasting.

Figure 6.4. The regular grooves and white ridges in the altocumulus are manifestations of buoyancy-induced oscillations in air pressure, known as gravity waves. There is little interaction and feedback between the gravity waves and the air current in which they form. Altocumulus lenticularis, photographed from Stratfield Mortimer, Berkshire, England, on July 1, 2007. © Stephen Burt.

Waves Are All Around Us

Figures 5.8 and 6.4 depict two very different types of wave motion; the common feature is the regularity of a spatial pattern. These patterns are described using fundamentally the same mathematics. This quantitative description of wave motion is familiar to physicists and mathematicians, and an intriguing question therefore presents itself: can we spot the ubiquity and diversity of wave motion by studying the governing equations of meteorology, as given in chapter 2? If we can, then are we able to use mathematics to focus on the behavior of those waves that are of importance to forecasters? The route to Rossby waves exploits the nearly linear behavior of the conservation law for total vorticity when evaluated near a very simple state of the atmosphere, such as the winds blowing uniformly to the east at higher altitudes.

It turns out that the technique Rossby employed can be generalized. In his quest to identify large-scale patterns, Rossby simplified his equation for the conservation of vorticity to arrive at a system that he was able to solve "with pencil and paper." More precisely, he assumed that a solution can be represented as the sum of a uniform simple flow pattern—a "basic state"—superimposed on which is a minor disturbance, or perturbation, say, in the pressure. By making this assumption and formulating it mathematically, solutions for the disturbances can be found from direct mathematical analysis without the need for powerful computational assistance. The crucial point is that the small disturbances do not influence the basic flow pattern—there is no feedback mechanism, as we usually find in nonlinear problems. Thus, different aspects of the flow patterns can be studied in isolation, without the complexity of the effects of the feedback processes that occur in real life.

At the beginning of chapter 4 we remarked that gentle ripples on the surface of a pond can be described very accurately by linear theory, but the breaking surf on the beach requires nonlinear equations. In the present context, we think of the gentle ripples on the surface of a pond as small disturbances about the state of rest of the pond, that of steady, calm, flat water. The ripples have very little influence throughout the depth of the water, and leave the water essentially undisturbed. On the other hand, the waves breaking on a beach are not small in comparison to the depth of the water on the shore, and all the water is in motion (typically moving the sand and pebbles as well). In this case, we cannot separate the flow into a part that is undisturbed and a part that we consider as a small disturbance.

We now wish to develop this concept more mathematically. When there is a simple, steady solution of the full problem, we may untangle and minimize nonlinear feedbacks in the system by applying the procedure just described. Here "steady" means that, at a given location, quantities such as the wind speed and direction are not changing in time. We then add gentle wave motions to the simple, steady solution. This whole procedure is called *perturbation theory*. The strategy involves making judicious approximations to the nonlinear equations with the result that the nonlinear problem is replaced by a series of simpler problems—which, as we explain in the following, turn out to be linear. Each addition of a gentle wave motion successively improves the results.

An example of a nonlinear problem whose solution may be found using perturbation theory is the swinging of a pendulum in the absence

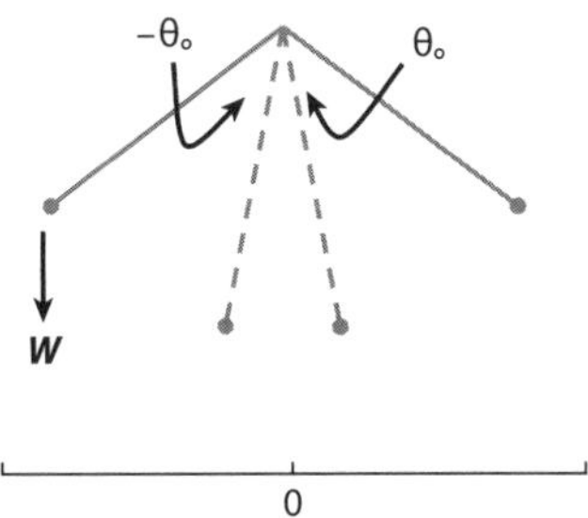

Figure 6.5. On the left is a photo of a grandfather clock showing the swinging pendulum in its glass case. Above is our schematic clock pendulum, whose position at different times is shown by dashed and solid lines (the solid lines show the extreme of a large amplitude oscillation). The weight, *W*, acts vertically downward on the swinging rod, tending to restore it to its stationary hanging position below the pivot—that is, above 0. But the pendulum swings past the vertical to the right. When there is no energy loss due to friction, then the pendulum starting at the angle on the left given by $-\theta_0$ comes to rest again at the angle on the right given by θ_0. This swinging motion recurs forever, forward and backward between the extreme angles θ_0 and $-\theta_0$. © masterrobert – Fotolia.com.

of friction. We introduced the double pendulum in chapter 4, and now we consider a single swinging rod. This is like the mechanism in a grandfather clock, as shown in figure 6.5. Newton's law tells us that the acceleration of the pendulum bob toward the vertical is caused by gravity trying to pull the mass downward, as we sketch in figure 6.5. In figure 6.6, the solid curve shows the value of the restoring force plotted in terms of the angle from the vertical. As the pendulum swings, it traces out this solid curve—to find out where it is at any given moment, we need a relationship between the angle from the vertical and the elapsed time.

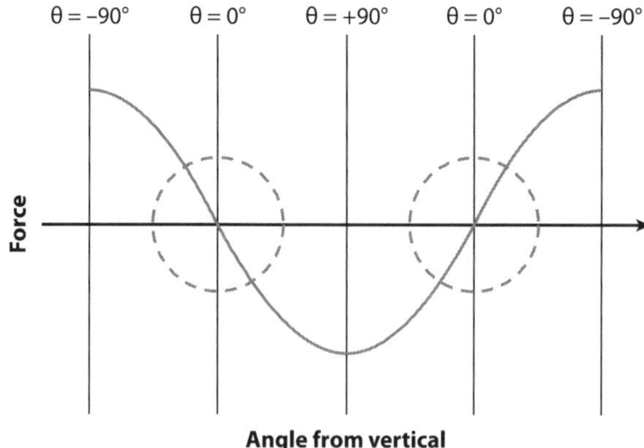

Figure 6.6. As the pendulum swings between the angles $-\theta_0$ and $+\theta_0$, where now $\theta_0 = 90°$, we show the restoring force (as the solid curve) in terms of the angle, θ, that the pendulum makes with the vertical (the values of θ shown along the horizontal arrow depict one complete cycle, to and fro). Inside the dashed circles we see the almost linear part of the solid curve.

The solid curve in figure 6.6 is a *sine curve*, described by the sine function of trigonometry. The differential equation that relates changes in the angle of the pendulum due to the gravitational force is a nonlinear equation. University science and engineering students soon discover that solving this equation to find the angle of the pendulum at any particular time in the future is actually quite difficult because the answer cannot be expressed in terms of simple mathematical functions.

However, if we look closely at the curve, we observe that when the angle is small—that is, when the pendulum is within only a few degrees of the vertical—then the relationship between the restoring force and the angle of displacement is almost a straight line, that is, almost *linear*. This is indicated in figure 6.6 by the portion of the solid curve within the dashed circles. So small oscillations of the pendulum, such as those indicated by the two dashed lines in figure 6.5, are approximated by a linear equation that can be solved both exactly and simply. Further, the swinging motion is predicted exactly forever. But is the actual nonlinear motion exactly predictable? Yes, despite the small nonlinear feedback, because the energy conservation law here is sufficient to control any chaotic motion in the pendulum behavior.

In chapter 4 we discussed the chaotic behavior that happens when a second pendulum is smoothly suspended from the first. There the energy conservation law could not control the feedback process sufficiently, and the "butterfly effect" of chaotic motion becomes possible.

That is, the future behavior becomes essentially unpredictable. Does our atmosphere behave more like a predictable single pendulum, or does it behave more like the unpredictable double pendulum?

A major success of this perturbation approach over the last few centuries has been the prediction of events recorded in astronomical calendars. When the Sun-Earth-Moon system is modeled to predict solar or lunar eclipses, can we neglect the influence of the other planets? In the case of the solar system, the Sun is the dominating influence, and we arrive at a remarkably good simulation of the motion of the planets around the Sun by ignoring the effects of the planets on each other by just using the gravitational field of the Sun in the equations. This approach can be justified on the grounds that the Sun is about one thousand times more massive than all the other planets put together, and therefore it dominates the gravitational interactions of the widely separated planets (just like a highly dominant protagonist in a group of people). Then we add in the gravitational correction due to each planet and begin a judicious perturbation calculation.

In the atmosphere, the recognition of the importance of the variation of the Coriolis force with latitude noted by Rossby (as discussed in chapter 5) in the large-scale motions of the gravity-dominated stratified air underpins his perturbation methods. The relative mass of the Sun versus that of the planets is so large that perturbation methods work exceedingly well in astronomy. In the atmosphere, the differing sizes of the restoring forces are not so clearly delineated, so interactions between the protagonists are much more subtle. Although perturbation methods give us valuable conceptual insights into complicated patterns of wind and rain, they cannot be used indiscriminately, nor can they be used *all* the time, nor do they yield predictions with astronomical accuracy.

The mathematics that predicts the future behavior of systems such as these swinging pendulum models is now known as *dynamical system theory*. The pendulum models have many applications in science and engineering that involve vibratory or wave motion. As we mentioned at the beginning of chapter 4, ripples on a pond can be described by linear theory in essentially the same way—the water in the pond sloshes up and down because gravity is tending to return the water to a dead calm in a way directly analogous to gravity restoring our swinging pendulum to its downward "calm" position. Applying perturbation theory to the equations governing the atmosphere when the air is near a state of rest allows us to identify oscillations in air pressure, which show up as waves in clouds when there is sufficient water vapor. It is possible to arrive at a formula known as a dispersion relation; Rossby's famous formula, given in tech box 5.1, is an example. The dispersion relation tells us how the propagation of a wave disturbance varies with the length of the wave. Other physical parameters of the problem may enter the relation. The key problem, eventually articulated by Charney and his colleagues after the meeting at Princeton, was how—out of all possible wave motions in the atmosphere—to focus on the cyclone waves, those of figure CI.7, that govern the larger scale weather.

Scales and Symphonies

Atmospheric motion is so much more complicated than pendulum or planetary motion. We use a visit to the seaside to illustrate another aspect of the problem Charney was grappling with. After sitting on a beach for several hours, we become aware of two very different types of motion in the sea. First, there are the fairly regular waves that break on the beach. But there is also the slow movement of the tidal water up (or down) the beach, which usually takes about six hours from fully out to fully in. We call the minute-by-minute behavior of the breaking waves "fast," and the hour-by-hour behavior of the tide "slow" (of course, there may be many other, usually tiny, waves on the beach). To predict the ebb and flow of the tide, we would not focus on the individual waves. Similarly, we want to identify fast and slow motions in the atmosphere because they play very different roles in forecasting the next day's weather.

The faster types of wave behavior can be observed in the forms of clouds as they pass overhead. Figure 6.4 shows the smaller, faster ripples in the clouds. These striations reveal the presence of certain rather special faster waves in the air pressure, the gravity waves. The slower, longer, more tidal waves are reflected in the motion of entire cloud layers, which extend over hundreds of kilometers. The sequence of ripples in the cloud shows ridges of uplifted air, and these are the signature of the faster waves. The uplifted air next subsides because it is denser than its surrounding air, and downward motion (usually) takes several minutes. When the air sinks, the clouds disappear as the cloud water vapor evaporates, only then to have sunk too far and to rise up again.

When planes fly through such rippling pressure patterns, the ride is usually described as turbulent.

The moving atmosphere, in addition to transporting air masses from place to place and from day to day, acts as a carrier of both sound and gravity waves. Computing these two forms of wave motion is much less important when it comes to predicting tomorrow's or next week's weather because we primarily need the hour-by-hour drift of the large-scale pressure pattern shown in the sweep of cloud in figure CI.7. The presence of sound and gravity waves in the equations for compressible fluid motion means that the mathematical and computer analysis becomes exceedingly subtle. Charney coined the term "meteorological noise" for such faster wave motions. In the qualitative theories of cyclones developed by the Bergen School, consideration of such motions did not enter the picture; conceptually they are easy to ignore. Removing them from accurate calculation is much more difficult and required ingenuity. Charney was to discover how to eliminate these waves from the mathematical analysis using judicious approximations to the governing equations.

In a letter dated February 12, 1947, to Lieutenant Philip Thompson at the Institute for Advanced Study, Charney discusses the problems of numerical forecasting based on the equations Richardson used, and the problems of singling out the motions in the atmosphere that are important to forecasters. He likens the atmosphere to a musical instrument on which we can play tunes. High notes are the sound waves and low notes are the more ponderous long waves that Rossby discovered. He claims that nature is a musician from the Beethoven school, rather than from the Chopin school, and only occasionally plays arpeggios in the treble and even then with a light hand. Charney remarks that we are familiar with the slow movements, and that only the academics at MIT and NYU are aware of the overtones.

Before we move on to Charney's breakthrough that paved the way for the first successful weather forecasts on a computer, we summarize his doctoral thesis—a seminal contribution to the systematic application of perturbation theory for simple waves on uniform basic flow states of the atmosphere. As we explained in chapter 5, conservation laws had shown how to begin to understand the great storm tracks along which weather cyclones travel, bringing weather across the Midwest in North America and across the North Atlantic to the shores of Western Europe. While the Rossby waves are essentially persistent features, the weather systems that follow these tracks are born, grow, and decay within a week or so. The quest to quantify the life cycle of these cyclones led to some of the great intellectual achievements of meteorological science in the twentieth century.

Following the discovery of the polar front by the Bergen School, the prevailing theories of the 1920s and '30s about cyclones were based on the preexisting discontinuity between the warm tropical air and the cold polar air (see figure 3.13). The front is a region in our atmosphere where small but significant changes in temperature-driven air pressure occur. We call a flow stable when these small changes do not grow; otherwise, it is unstable. What is needed is a theory that builds on the cruder approximations made by Rossby, and that couples the changes in the winds with the thermal and moisture processes in such a way as to create the potential for such instability that a cyclone could develop from an initially small pressure perturbation.

In contrast, ever-increasing observations of the upper atmosphere in the 1930s showed, in the words of Charney (1947), that "the number of surface frontal perturbations greatly exceeds the relative small number of major waves and vortices at upper levels." He goes on "it is natural . . . to attempt to explain the motion of the long waves in terms of the properties of the general westerly flow without reference to frontal surfaces." These factors helped lead meteorological thinking in the 1940s away from the lower-level polar front and toward the higher-level westerlies in the middle to upper atmosphere. It was recognized from more extensive wind measurements that these winds increase in strength with height in a roughly linear fashion. When coupled with the equator-to-pole temperature gradient, the resulting basic state was a more promising starting point from which to study the triggering mechanisms responsible for the formation of the most familiar of the midlatitude weather systems—the ubiquitous cyclone.

As so often happens in science, two attacks on the problem were made at similar times but in widely separated places. In the 1940s two very different people had produced two very different answers to this birth of cyclones question. In the United States, as discussed earlier, Charney published in 1947, and in the United Kingdom Eric Eady published in 1949. Charney's paper is widely regarded as one of the most important papers in twentieth-century theoretical meteorology because

Charney changed our culture by changing our attitudes to mathematical procedures on the fundamental equations. He did this by showing that the more intuitive results of Rossby could be systematically (mathematically) squeezed out of the laws of meteorology. This 1947 paper encouraged others to be more mathematically rigorous and introduced analysis of the equations as a valid mode of demonstration alongside the more traditional analysis of atmospheric data.

Charney's 1947 paper itself is still not an easy read—it is a tour de force in terms of mathematical analysis of the problem. The paper, entitled "The Dynamics of Long Waves in a Baroclinic Westerly Current," was published in October 1947 in the *Journal of Meteorology* (now known as the *Journal of Atmospheric Sciences*). The young Charney holds nothing back in the introduction to his paper when he states, "The large-scale weather phenomena in the extra-tropical zones of the earth are associated with great migratory vortices (cyclones) travelling in a belt of prevailing westerly winds. One of the fundamental problems in theoretical meteorology has been the explanation of the origin and development of these cyclones." Near the end of the introduction he states, "This work presents a clear physical explanation of instability in the westerlies and establishes necessary criteria . . . that any exact mathematical treatment of baroclinic waves must satisfy." These *baroclinic waves* are so called because of the dependence of pressure on both density and temperature: Charney called them "cyclone waves."

Eady's approach is mathematically and conceptually more straightforward and has become an essential topic for all undergraduate courses in modern meteorology. The Eady wave, as his solution is now known, is one of the required intellectual stepping-stones for students of the subject. Eady alluded to an interesting analogy borrowed from the Darwinian theory of evolution to explain his ideas. He imagined that the wavy perturbations with their slightly different horizontal scales were competing with each other to survive. The ones with the maximum growth rate should quickly dominate the others, and so become the most probable development. Eady states, "Were it not that 'natural selection' is a very real process, weather systems would be much more variable in size, structure, and behaviour." This theory of natural selection suggests that the dominant disturbances should have a wavelength of some four thousand kilometers, and these are now regarded as the embryonic cyclones

Figure 6.7. Eric Eady (1915–66) graduated from Christ's College, Cambridge, with a BA in mathematics in 1936 and joined the U.K. Met Office as a technical officer in 1937. During the Second World War, meteorology was both a profession and a hobby for Eady. Working with no advice or encouragement, he used his spare time to advance the subject that was at the heart of his daily duties. By the end of the war, Eady decided to devote himself to theoretical meteorology, and after extending his earlier research, he obtained a PhD from Imperial College London in 1948. He had already come to the attention of Jack Bjerknes, who had invited him to Bergen in 1947, and invitations to visit von Neumann in Princeton and Rossby in Stockholm followed shortly after. Only a small fraction of his work has ever been published. In the obituary notice in the *Quarterly Journal of the Royal Meteorological Society*, his paper "Long Waves and Cyclone Waves" is described as "a masterly summary of his early work, displaying it in a clear and elegant manner as the logical extension of the physical hydrodynamics of V. Bjerknes and his school, reinforced with the physical awareness so earnestly demanded by that pioneer, and brought close to the ultimate that can be achieved by a formal analytical approach." Courtesy of Norman Phillips. Reprinted with permission.

outside the tropics. They are "embryonic" because the theory developed by Charney and Eady is linear.

Charney and Eady were solving the first step in a perturbation procedure that would cut the feedback knot, and so the theory can only describe the smaller amplitude wavy disturbances. The linear theory is no longer valid when the cyclone becomes recognizable as in figure CI.7, which is a highly nonlinear state. However, in contrast to Rossby's earlier work, the basic state of the atmosphere used by both Eady and Charney was more realistic. The basic state that they chose to perturb involved the strengthening westerlies in the middle atmosphere and the decrease of temperature toward the pole. Both of these could provide the energy needed for growing disturbances, and sufficiently so that their solutions recognizably modeled the growth of idealized cyclones.

Charney soon followed his groundbreaking doctoral work on the development of cyclones with another seminal contribution to meteorology, inspired by von Neumann's project to compute the weather. In a paper published in 1948 in the journal *Geofysiske publikasjoner* (*Geophysical Publications*) entitled "On the Scale of Atmospheric Motions," he lays out the basic theory that was to revolutionize theoretical meteorology for the next fifty years by clearing some of the hitherto impenetrable

Tech Box 6.1. The Long Wave Problem: Eady's Paradigm

Eady's model consists of a basic zonal flow, U, to the east with a decreasing temperature gradient to the north (simulating the equator-to-pole temperature contrast in the zone of westerly winds). By considering a limited zone in temperate midlatitudes, Eady ignored Rossby's variation in the Coriolis effect. The wind changes with height, and this balances the horizontal temperature gradient, a result known as the thermal wind relation. Eady then assumed there was a "rigid lid" at the top of his model atmosphere, which simplified the effect of the stratosphere. At the heart of his model lay the more realistic basic state, to which he added wavy perturbations of the form

$$A(z)\ \exp[ik(x-ct)]\sin(\pi y/L).$$

Here x and y are the eastward and northward coordinates, respectively, k is the zonal wave number, c the phase speed, $i = \sqrt{(-1)}$, L is a horizontal length scale, and $A(z)$ is an unknown amplitude depending on the height z. Substituting this into the governing equations, he found that under certain conditions the phase becomes *imaginary*, and therefore the product $-ic$ yields exponential growth in time—this is the hallmark of instability. The results of these calculations compare well with computer simulations, such as those shown in figure 6.8.

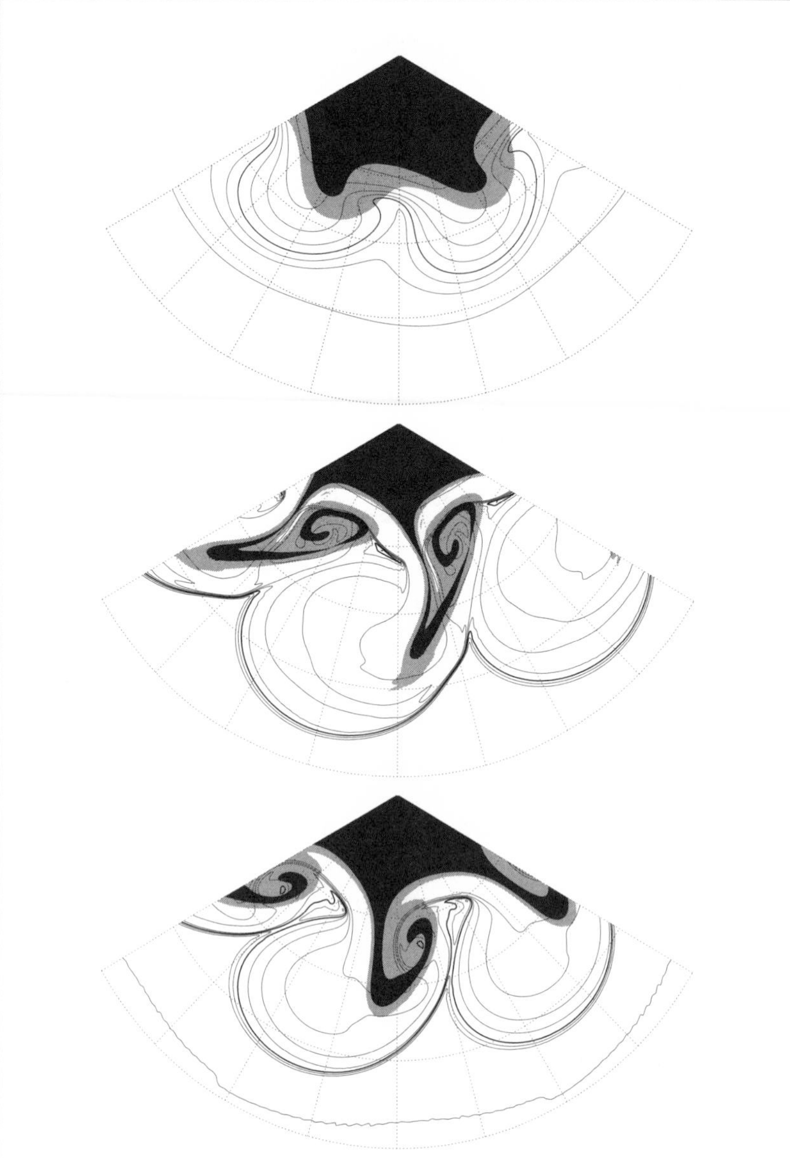

Figure 6.8. A modern computer simulation of a northern Atlantic air flow developing over an eight-day period. The first picture shows the flow four days after the initial linearized state of tech box 6.1, while the second picture shows the strong cold and warm fronts that have formed by day six as well as the beginning of an occlusion. The shading depicts potential vorticity at a particular height in the atmosphere, and the contours show the potential temperature (as defined in tech box 5.2). © John Methven.

intellectual fog created by nonlinearity. His introduction is once again decisively clear (albeit with a somewhat modest first few words):

> In a recent publication entitled *The Dynamics of Long Waves in a Baroclinic Westerly Current* the writer pointed out that, in the study of atmospheric wave motion, the problem of integration is greatly complicated by the simultaneous existence of a discrete set of wave motions all of which satisfy the conditions of the problem. . . . Whereas only the long inertially-propagated waves are important for the study of large-scale weather phenomena, one is forced by the generality of the equations of motion to contend with each of the theoretically possible wave types. . . . This extreme generality whereby the equations of motion apply to the entire spectrum of possible motions—to sound waves as well as to cyclone waves—constitutes a serious defect of the equations from the meteorological point-of-view. . . . It means that the investigator must take into account modifications to the large-scale motions of the atmosphere which are of little meteorological importance and which only serve to make the integration of the equations a virtual impossibility.

Charney sets out to rid us of the encumbrance of what he called "meteorological noise"—the mass of irrelevant small detail in the pressure field, some of which appears in the small-scale patterns in the clouds shown in the photos in figures CI.7 and CI.8 (in the color section).

Charney's intuition was to seek a basic state and associated wave motion that is in approximate *geostrophic balance*—the balance between the dominating horizontal pressure gradient and Coriolis forces that explains Buys-Ballot's law. This was a significant step: hitherto, perturbation methods had been applied to basic states that were constant in space and time. Now Charney wanted to introduce a basic state that evolved slowly and possessed spatial structure. The outcome of such a strategy is a new starting point, a system of nonlinear equations, so that straightforward linearization is generalized to a more realistic basic flow state. The procedure yields a model that is still relatively simple and is amenable to analysis with pencil and paper.

The terms in the basic equations for large-scale atmospheric motions typically have quite different magnitudes—the effects due to the rotation of the Earth (the Coriolis effect) and the horizontal pressure-gradient force are nearly equal and, roughly, about ten times greater than the horizontal accelerations (see figure 6.9). Charney showed this ranking of the typical sizes of the major forces in his 1948 paper.

Geostrophic balance, the approximate balance of these leading terms in the horizontal force equations, dominates most of the dynamics of weather systems. Hence, we need only approximate the smaller acceleration terms. Because these acceleration terms are the source of much of the nonlinearity and associated analytical difficulties (as explained in chapter 4), this approximation cuts the nonlinear feedback to manageable proportions. However, such simplifications should respect the importance of these controlling influences. That is, instead of using the horizontal Newtonian laws to predict the change in the wind, Charney proposed using them to predict the geostrophic wind itself, and then to correct for the smaller variations due to the smaller accelerations, thus weakening the Gordian knot. So what Charney was proposing in this second paper was much more ambitious: the simple westerly winds of the basic state of his first paper were now replaced by (nonlinear) geostrophic winds, and then a perturbation technique calculates the fluctuations about the geostrophic state.

One of the highlights of Charney's 1948 paper, about half way through the text, is the statement that "the motion of large-scale atmospheric disturbances is governed by the laws of conservation of potential temperature and absolute potential vorticity, and by the conditions that the horizontal velocity be quasi-geostrophic and the pressure quasi-hydrostatic." Such a view is now called *quasi-geostrophic* (or QG) *theory*. It has been said by various meteorologists, perhaps over the occasional

$$\frac{\text{horizontal acceleration}}{\text{horizontal coriolis force}} \sim \frac{10/10^6}{10^{-4}} \sim \frac{1}{10},$$

$$\frac{\text{vertical acceleration}}{\text{acceleration of gravity}} \sim \frac{CW}{gS} \lesssim \frac{C^2H}{gS^2} \sim \frac{10^2 \times 10^4}{10 \times 10^{12}} \sim 10^{-7},$$

Figure 6.9. The top equation is Charney's calculation, from his 1948 paper, of the ratio of the typical magnitude of the horizontal acceleration to the Coriolis effect. The Coriolis effect dominates by about a factor of 10, thus justifying geostrophic balance (Buys-Ballot's law as understood by Ferrel). The bottom equation is Charney's calculation of a ratio in the vertical, thereby justifying hydrostatic balance. Notice that the typical vertical acceleration is much, much smaller than the acceleration due to gravity. *The Atmosphere: A Challenge.* © American Meteorological Society. Reprinted with permission.

drink in the conference bar, that most of what is now understood about the dynamics of the atmosphere and oceans still comes from studying problems using QG theory. When we consider the intractability of Euler's equations—even without the additional complications of heat and moisture processes—then this quote from Charney's paper must rank among the most effective and profound statements of modern meteorology. His remark also underlines the importance of the role of conservation laws and the importance of being able to make judicious approximations. Charney had well understood the anatomy of the problem: using his musical analogy, he had identified the different sections of the orchestra and understood the rules for creating a pleasing harmony by balancing these sections even though the music he made was dominated by the bass section.

Charney's 1948 paper established a mathematical basis for the description of large-scale motions in the atmosphere. He had finally got enough realistic weather into the basic state of the atmosphere that QG theory became the touchstone for the remainder of the twentieth century. Charney had assimilated nearly a century's worth of dynamical meteorology, which highlighted the importance of large-scale circulations on smaller-scale weather, and had succeeded in describing it for the first time in a mathematically predictive way. In a sense, he produced a mathematical description of what a forecaster is trained to see in a chart—a very smoothed-out view of weather. He was able to eliminate from the general solution the motions that forecasters usually instinctively ignore—the sound waves, gravity waves, and much more detail.

Geostrophic balance is characterized by horizontal flow around the pressure contours, and it explains the prevailing winds in developing cyclones. The overall motion of the cyclone can be understood in terms of the conservation of potential vorticity on the cyclone scale. Thus, Rossby waves explain the time-averaged tracks of these weather systems around the polar front, and QG theory is needed to explain the development of cyclones that arise from instabilities in flows with horizontal temperature gradients. The preceding discussion amounts to a remarkable top-down view of the key processes that need to be represented in order to produce accurate weather forecasts. Moreover, these concepts, which correspond to some very strong constraints on the motion of the atmosphere in the presence of heat and moisture processes, are effectively buried beneath the complexity of the full nonlinear equations we use in weather forecasting. How, then, do we ensure that the automatic computations of billions of bits of detail all add up in just the right way—by respecting the principles and constraints just mentioned—so that the correct "big picture" of next week's weather is found rather than a turbulent, chaotic mess that the nonlinearity quite often allows? This issue will be pursued in the remaining chapters.

The clarity and depth of Charney's 1948 paper were consequences of the insights he gained during his doctoral research, his research dialogue with Rossby at the University of Chicago, and through his engagement with von Neumann's project. Charney had received a scholarship to travel to Stockholm to make contact with European meteorologists after the Second World War. He then visited Chicago to explain his thesis work to Rossby and the Chicago research team. When Charney and Rossby found an instinctive empathy and creative understanding, they talked virtually every day for eight months, after Charney delayed his trip to Europe. This extended interaction with Rossby was to change his life—and it initiated modern weather forecasting when Charney was asked to head the Princeton project.

Analyzing wave motion using rigorous methods of perturbation theory about a simple basic state is the cornerstone of Charney's doctoral thesis. Perturbation theory tells us whether a wave-like disturbance will remain small and wave-like or whether it will grow. In the case of growing oscillations, there comes a point when the linear theory that identified these modes becomes invalid, but the linear theory does tell us under what circumstances waves will grow in size. Charney's 1948 paper had shown how to successfully perturb a more complicated basic state, one that allowed feedback processes, using the geostrophic winds. While we would be forgiven for believing this might be a retrogressive step—after all, nonlinear feedback is precisely the phenomenon we are trying to vanquish—Charney incorporated the notion of scale into his equations so that the nonlinearity was retained just where it was needed to give more realistic cyclone weather descriptions: in the conservation laws and balances that control the evolution of cyclones and fronts. In chapter 5 we suggested that Rossby had taken a razor to the governing equations in order to cut away everything except the terms required to describe large-scale wave motion; here we might liken Charney to a surgeon, using a scalpel to remove just the troublesome waves—a more delicate and subtle operation.

The hydrostatic and geostrophic models of Charney and Eady, which marked the first great transformation of meteorology in the late 1940s, provided a foundation for the second great transformation: the production of numerical forecasts on the very first electronic computers.

Second Time Around

The enormous accomplishment of Charney in transforming theoretical meteorology was motivated in the first instance by the questions posed at the Princeton Conference in Meteorology in 1946. Later, in 1950, a highly simplified practical model of weather was created that was amenable to numerical integration via finite differences and a fixed grid. The other big difference compared to Richardson's calculation was the use of the new electronic computer.

By the end of the conference the group still did not know how to avoid the pitfalls of Richardson's attempt. It was noted that Charney made some rather abstruse comments about the need to filter the gravity waves that had contributed to the failure of Richardson's forecast, but no one really paid much attention. It was not until a year later that Charney, as indicated in his 1947 correspondence with Thompson, began to formulate a general method for filtering the "meteorological noise," focusing the rather modest computer power on predicting the "deep chords" and ignoring the "rapid lighter scales" of the faster processes. Rossby was still advocating the use of the full equations similar to those Richardson had used, and, according to Charney (in an interview published in the Charney memorial volume), Rossby had not yet considered using a much simpler model that filtered the high-frequency noise.

Indeed, Charney's engagement with the new computer project led him to formulate, while on a year's leave in Stockholm and Bergen, the ideas we discussed in the previous section. Charney said that the ideas came to him after months of undisturbed thinking. To add to the triumph of his doctoral thesis, Charney was appointed to lead the team at Princeton. This was a wise decision on the part of von Neumann, who also invited Eady and Arnt Eliassen (recognized as one of the great meteorologists of his time) to visit and join the team. Charney, Rossby, and von Neumann were very different characters—with different ages and at varying stages in their very different careers—but they all shared a similar apprenticeship as mathematical physicists. The success that the team was to achieve was due in no small way to their ability to communicate ideas and to draw on each other's strengths.

The team did not need to be reminded about the difficulties in accurately measuring both the departure of motions from geostrophic balance and the horizontal convergence of air motion. As we explained in chapter 3, the difficulties of accurate measurements make it very difficult to control errors that occur when solving the equations used by Richardson. Charney had by that time realized that the errors in the computer calculations of the pressure patterns would usually only initially manifest themselves in the form of gravity waves, but this would be enough to lead to the sort of catastrophic error that Richardson's calculation had produced. Finding a solution to this problem was as important as having the machine available to perform the time-consuming calculations.

Whereas Richardson had tried to include virtually every effect into his model, Charney realized that simplicity was essential. Steps must be taken to control the errors introduced as a result of the high-frequency gravity waves that masked the required low-frequency (tidally drifting) forecasting signal in the pressure and wind fields. Powerful computers alone would not circumvent that problem; moreover, a model based on Richardson's equations would far exceed the computing technology available to them.

Charney realized that he had to simplify the QG theory a little more before it was in a form amenable to the computer experiment they envisaged. In his final analysis, Charney eventually concluded that the equation they were being driven toward was none other than Rossby's equation for the conservation of total vorticity, the one we discussed in chapter 5. He did not reach this conclusion until about 1949, several years after he had sat next to Rossby at the conference in Princeton where they had both pondered the problem of how to omit or filter out the gravity waves. In using Rossby's equation for the conservation of total vorticity, vertical motion was ignored in the prediction, which was a radical approximation because vertical motion is perhaps the most critical factor behind phenomena such as the energy conversion required to power the cyclone waves, cloud formation, snow, and rainfall. So Charney's and Rossby's theories gave forecasters the means to forecast changing pressure patterns over limited time periods of one to two days but nothing more. But it would be a start!

The essential tool in the project, the Electronic Numerical Integrator and Computer, or ENIAC for short, was the first public multipurpose, electronic, digital computing machine, both in design and construction. It became operational in December 1945. The machine was massive in the old-fashioned sense—it was bigger than a bus and had less than one millionth of the processing power of a modern mobile phone! Historically, the first programmable computer was the highly secret Colossus at Bletchley Park in England, which was used to break the Enigma code in the Second World War, but it was not a multipurpose machine.

ENIAC contained eighteen thousand vacuum tubes (or valves with the appearance and the size of light bulbs), seventy thousand resistors, ten thousand capacitors, and six thousand switches. Its forty-two main panels were arranged along three walls of a large room and consumed 140 kw of power. A modern personal computer generates heat, and several warm a room when left switched on. ENIAC consumed the equivalent power of nearly five hundred desktop personal computers. An essential adjunct to the ENIAC itself was the punch-card equipment that served as the large capacity read/write memory. The reading was

Figure 6.10. ENIAC at the Moore School of Electrical Engineering, University of Pennsylvania. U.S. Army photo.

performed by card readers and writing by punch card. These punch cards were like old-fashioned stiff paper library reference cards except that they had holes punched in them to store data in a coded form. Typically a box of such cards provided the data input to the computer; and they had to be in the right order. Woe to a careless programmer who spilled a box on the floor and had to correctly reorder all the cards.

ENIAC could add two numbers in about 0.2 milliseconds and multiply two numbers in 2 to 3 milliseconds. The computer was designed as a "hard-wired" machine. That is, rather like the old plug switchboards that were used to control telephone exchanges and communications, so programs that controlled computations had to be "wired in" to ENIAC. This made programming very cumbersome and time-consuming. The stored program, which is the basis of the design of computers as we know them today, made its first appearance about the time ENIAC went operational. A stored program is one in which individual commands can be stored as numbers in direct-access memory. It is constructed from a repertoire of commands that is universal in the sense that the solution to any problem that can be solved by computer can be solved by using these and only these commands.

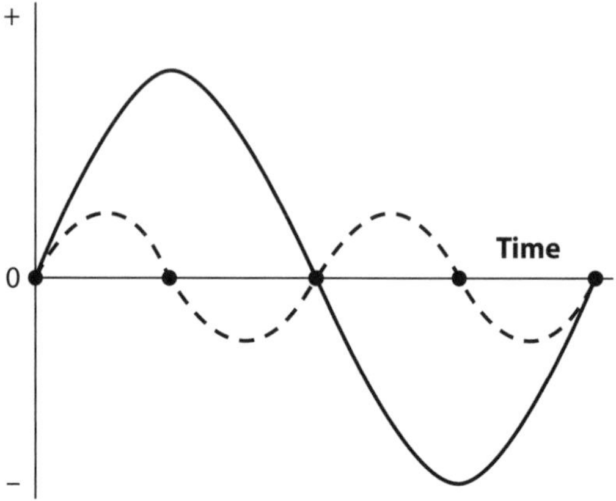

Figure 6.11. The solid curve shows a wave plotted over a distance of one complete wavelength. The dashed curve shows a faster wave with half the wavelength and twice the frequency. When we represent the slower wave with computer pixels, we need at least five pixels, shown by the solid circles: at the peak, at the trough, and at the three locations where it crosses the axis (where the wave takes the value 0). When we represent these two waves on these pixels, we just "see" the slower wave. Note that the pixel values are zero for the shorter, faster wave, and so we cannot "see it." We would need at least twice as many pixels to see the faster wave—but then would miss an even faster wave.

A paper by Charney, von Neumann, and another Norwegian meteorologist, Ragnar Fjørtoft, published in *Tellus* in November 1950, describes the first major results of the ENIAC project. Because each time step forward was at least one hour, and the steps in space about eight hundred kilometers in the east–west direction, only the simplest and slowest types of motion could be computed. Even this took a week of computer time. This is why Charney and his team had to find an extremely simple model that only had these types of solutions. Figure 6.11 illustrates the difficulty of representing fine detail on coarse grids.

How does the numerical model that predicts the changing pressure patterns work? The idea is beautifully simple. First, because the equation of motion is the conservation law for total vorticity, only the rotational motion of the wind is used in the computation, which is a reasonable first approximation in the midtroposphere. The convergence and divergence of most wind patterns takes place above (at the cirrus cloud level) and below (from the ground to cumulus cloud level), as we sketch in figure 6.12.

In accordance with the ideas we developed in chapter 5, we think of the atmosphere as a single thin shell of air (as shown in figure CI.5). The lower boundary is taken to be the Earth's surface and the upper boundary the tropopause, above which lies the stratosphere. The troposphere,

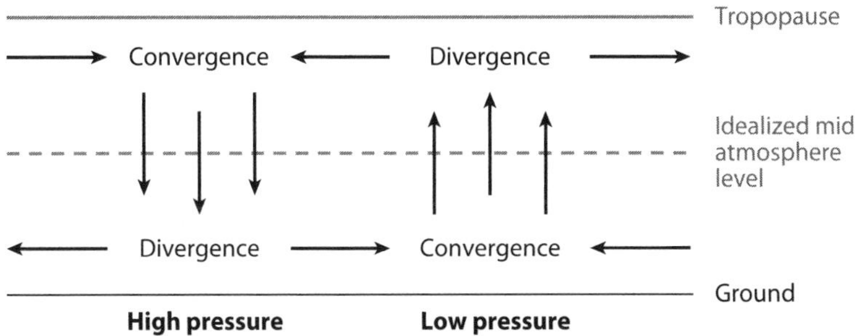

Figure 6.12. This is a typical large-scale wind pattern in a cross section of the atmosphere that extends to about nine kilometers above the ground and is several thousand kilometers in horizontal extent. At the lower level (say, ground to three kilometers) we see mainly cumulus clouds, while at the higher level (above six kilometers, say) we might see a type of cirrus cloud, depending on the moisture present. The computer calculations were at the idealized midlevel of about five hundred millibars, where we show both upward and downward vertical motion. However, no attempt was made to directly calculate this vertical motion, and it was assumed that the convergence exactly vanished at this midlevel.

which is the region between these two boundaries, is where most of our weather occurs. We predict certain airflows by simulating only the vorticity in the middle of this layer. That is, by referring to figure 6.12, between the ground and the tropopause will be a surface on which the convergence is approximately zero—where the air moves with very little horizontal "squashing in." Most of the convergence and divergence occurs either above or below this midlevel surface. Nearer to the sea level and the tropopause, the airflow is described by a combination of vorticity and convergence. The change in the sea-level pressure is a consequence of converging and diverging airflow moving over the ground, which we recognize as highs and lows. The solutions that we seek to compute are smoothed-out versions of the picture in figure CI.7, where all the detail apart from the large swirling airflow is ignored.

Weather forecasters made their predictions from maps of the pressure patterns at ground level, so a way of relating the middle atmosphere pressure to that at sea level was required. Bjerknes himself had drawn maps of "height" to achieve this. This height is actually the height of the 500 mb surface, here shown in figure 6.13. So, one of the key variables that we want to predict is known, in meteorological parlance, simply as "height." Instead of showing variations in pressure at the Earth's surface or at a constant altitude, meteorological charts depict the height of a constant pressure surface, an *isobaric surface*, above sea level (see figure 6.13).

In regions where there are no horizontal variations in pressure, constant height surfaces and isobaric surfaces are parallel. But, because of the changes in air density, a surface of constant pressure rises in warm, less-dense air. We may imagine heating the air in a balloon and

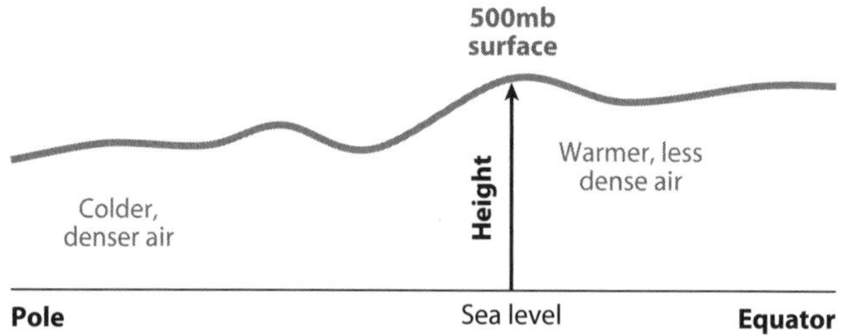

Figure 6.13. The curve indicates the five hundred millibar pressure surface. Its elevation above sea level is known as "height," which varies depending on the density (and, hence, temperature) of the air beneath it. This surface is typically around five kilometers above sea level.

watching it expand, so a heated air layer also expands and its upper surface rises. Similarly, the upper surface lowers in colder, denser air. This transformation between height and pressure means that we can use pressure as a vertical coordinate (that is, as a measure of altitude), which is common practice in meteorology and in some numerical weather prediction models.

To get an intuitive feel for how the vorticity is related to the height field, we think of the map of the height of the 500 mb pressure surface as like a map showing height above sea level in a mountainous region (see figures 6.13 and 6.14). We know that the closer the contours are on a hiking map, the steeper the terrain; in a similar way, we also know that the closer together the height contours are on the weather map, the stronger the wind will be. If we now imagine ourselves standing on the top of a "hill" in this landscape that represents an undulating constant-pressure surface, then the vorticity of the flow is related to the rate at which the *slope* of the landscape changes. That is, the vorticity on this pressure surface is proportional to the surface curvature.

At the heart of the mathematical model used in the ENIAC project was the equation for the conservation of total vorticity. In this model, the winds are in geostrophic balance, which means that they blow parallel to the height contours, as shown in figure 6.14. In a static atmosphere, the pressure, temperature, and density vary in a prescribed way with height. For a sequence of pressure values (for example 100 mb, 200 mb, . . . 1,000 mb), we construct each height surface by repeating the procedure shown in figure 6.13. Up to this point we have considered the weather pixels to be located at different elevations in the atmosphere, where elevation is measured in terms of distance. The new construction enables us to take the radical step and transform our view to one in which the elevation of the pixels is measured by pressure surfaces. So, our sequence of pressure values (100 mb, 200 mb, . . . 1,000 mb) becomes the new vertical coordinate (with 1,000 mb at sea level and, say, 100 mb at the top of our atmosphere).

The ENIAC numerical forecast now proceeds as follows. Suppose we have the 500 mb height chart shown in figure 6.14, which depicts the synoptic situation at 12:00 noon on a particular day. We place a grid of pixels over the chart and by a process of interpolation (that is, working in from the corners of the squares into where the actual contours are), we read off the contour height at each point on our grid. When this is

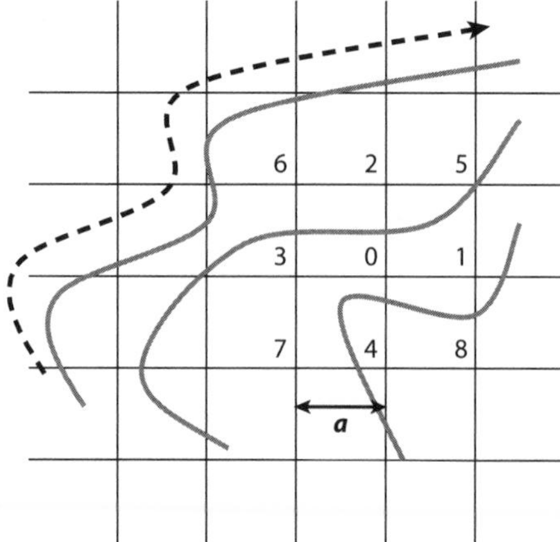

Figure 6.14. The contours of height are shown as solid curves and represent values of the elevation of the 500 mb surface. We show the geostrophic wind as the dashed curve, blowing parallel to the leftmost contour. The corners of the pixels are commonly referred to as grid points, and these are labeled by numbers. We estimate or interpolate the height at each grid point by the following procedure. Consider the contour that runs between grid points 1 and 8, at a height 5,500 meters, and the contour that runs between the points 0 and 2, at a height of 5,300 meters: then a value somewhere between 5,500 and 5,300 will be interpolated to grid point 1. This is done for all the points over the domain.

completed by following the procedure described in figure 6.14, we have estimated the value of the height field at each discrete point on the grid of pixels. That is, we have converted the continuous contoured field into a finite number of discrete values. The difference in the values of the height from one grid point to its next nearest neighbor gives the derivative of the height in that direction over the domain, and from these derivatives we can calculate the geostrophic wind. In technical terms, the geostrophic wind in the east–west direction at point 0 is proportional to the difference between the values of the height at points 4 and 2, divided by twice the separation of the points, $2a$.

The conservation law states that the contours of total vorticity are transported by the geostrophic wind. The vorticity is computed from the geostrophic winds by a very simple arithmetic sum. Referring to figure 6.15, the vorticity at point 0 is found by adding the values of the height field at points 1, 2, 3, and 4, subtracting four times the value of the field at point 0, and dividing by the square of the grid separation, $a \times a$. Thus, vorticity is zero when each grid value is the exact average of its nearest four neighbors.

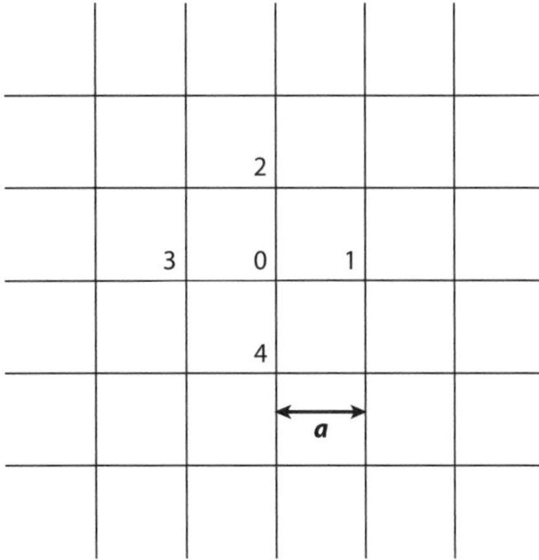

Figure 6.15. We show a local numbering of the grid for values of vorticity. The vorticity at 0 is the sum of the heights at 1, 2, 3, and 4, minus four times the value of the height at 0, and divided by $a \times a$. So vorticity is a measure of the deviation of the height at 0 from the average of that at its nearest neighbors.

To each grid-point value of the vorticity we add the value of the Coriolis parameter at that point. Thus, in the language of chapter 5, we find the total vorticity as the sum of the local and planetary vorticities. These calculations are performed at each grid point over the domain, and then these values are plotted on a second grid. Using the second grid we can now calculate the derivatives (or rates of change) of the vorticity; these are needed for the vorticity transport term in the equation. Once again, these are calculated using the finite differences of the vorticities at each grid point. Because we are working on a grid with boundaries, as Richardson did, we have to make some approximations when calculating the derivatives at the edges of the grid.

The equation we solve is of a special type that allows us to extend values smoothly from the grid boundary to the grid interior. It follows that if we know the values of the rate of change of height with time at the boundaries of the region, then the corresponding values can be calculated at all interior points. If we take a time step of, say, one hour, then the height field is computed at, say, 1:00 p.m. by adding the derivative (at the rate of one hour) that we have just computed to the value of the height field at 12:00 noon. By repeating the whole operation, we produce a forecast step by step according to the following algorithm:

height at 1300 = height at 1200 + total hourly change in height with time at 1200.

So the basic numerical methods followed Richardson's ideas, but the model was much simpler, and a machine did the calculations by systematically working over all the grid points at each time step.

The pressure field is obtained from the above-calculated vorticity by integrating what is known as the *balance equation*. This equation tells us how the height field is related to the vorticity; thus, we obtain the winds that move the patterns around from the height by the geostrophic wind law as in figure 6.14. From the wind fields we calculate pattern movement; we then repeat the whole process over and over again for successive time steps. This creates the forecast.

It's Snowing in Washington!

ENIAC performed all the arithmetic involved in the simulations, but all the preparation for the arithmetic—the input to and output from these calculations—involved manual work by the human operators who produced the punched cards and fed them into the machine. The first program, designed by Charney and von Neumann, required not only the preparation of instructions for ENIAC but also instructions for the small army of operators. There were a total of sixteen operations that together formed a single time step of the simulation. Six of the operations were carried out by ENIAC, and ten were punch-card operations. Each ENIAC operation produced output via the cardpunch, and the punch cards had to be collated and sorted to produce input to subsequent operations. Von Neumann ingeniously contrived the intimate coupling between the punch-card operations of the human operators and ENIAC calculations.

Once again, Platzman's historical review captures the story perfectly:

> On the first Sunday of March, 1950 an eager band of five meteorologists [a stark contrast to the 64,000 envisioned by Richardson] arrived in Aberdeen, Maryland, to play their roles in a remarkable exploit. On a contracted time scale the groundwork for this event had been laid in Princeton in a mere two to three years, but in another sense what took place was the enactment of a vision foretold by L. F. Richardson . . . [about 40] years before. The proceedings in Aberdeen began at 12 p.m. Sunday, March 5, 1950 and continued 24 hours a day for 33 days and nights, with only brief interruptions. The script for this lengthy performance was written by John von Neumann and by Jule Charney.

Their goal was a twenty-four-hour forecast for North America for February 14, 1949. Quite naturally, the team had decided to test their ideas by attempting to reproduce a known event in which the initial conditions for the forecast, together with knowledge of what the answer should be, were well documented. By the end of the thirteenth day of the operations, a twelve-hour forecast had been made, and by the end of their allotted time on the machine, four twenty-four-hour forecasts had been made, in particular one for February 14 and another one for January 31, 1949. The computation time (that is, the actual time the computer was running on this project) for these forecasts was almost exactly the same as the time it had actually taken for the real motions to take place, which was twenty-four hours. The human input was enormous, but the results were very encouraging.

The prediction for the forecast for the next day was first made by twenty-four steps of one hour, but this was changed to twelve steps of two hours, and then eight steps of three hours, when it was realized that the results were virtually the same. This was the bonus that arose from using a program that focused only on the cyclone movement—Charney's lowest notes of the atmosphere: by using only eight steps instead of twenty-four, the process became three times faster and only took eleven days to actually compute on ENIAC.

The area covered by the ENIAC forecasts was North America, the Atlantic Ocean, and a small part of Europe, together with the eastern part of the Pacific Ocean. Because it was more difficult to obtain observations over the oceans, and because an ocean lay on the western boundary of their domain, the team was very aware of the errors that this situation might create. The data that they used to start their forecast was obtained by manually interpolating "data" from the U.S. Weather Bureau's operational charts for 0300 GMT (3:00 Greenwich Mean Time) on the days that forecasts were made: January 5, 30, and 31, and February 13, 1949. On these days the weather was dominated by very large-scale features, which they hoped to simulate successfully. We give the results of their calculations for January 5, 1949, in figure 6.16.

Even though the contour patterns for the forecast and for the actual weather are roughly similar, the ENIAC forecast for January 5 was poor. The system of interest was an intense low-pressure system over the United States, and they failed to predict the correct amount by which this system moved over the continent, and the shape of the depression was also distorted. On the other hand, the forecast for January 30 contained a number of very encouraging features. The displacement and

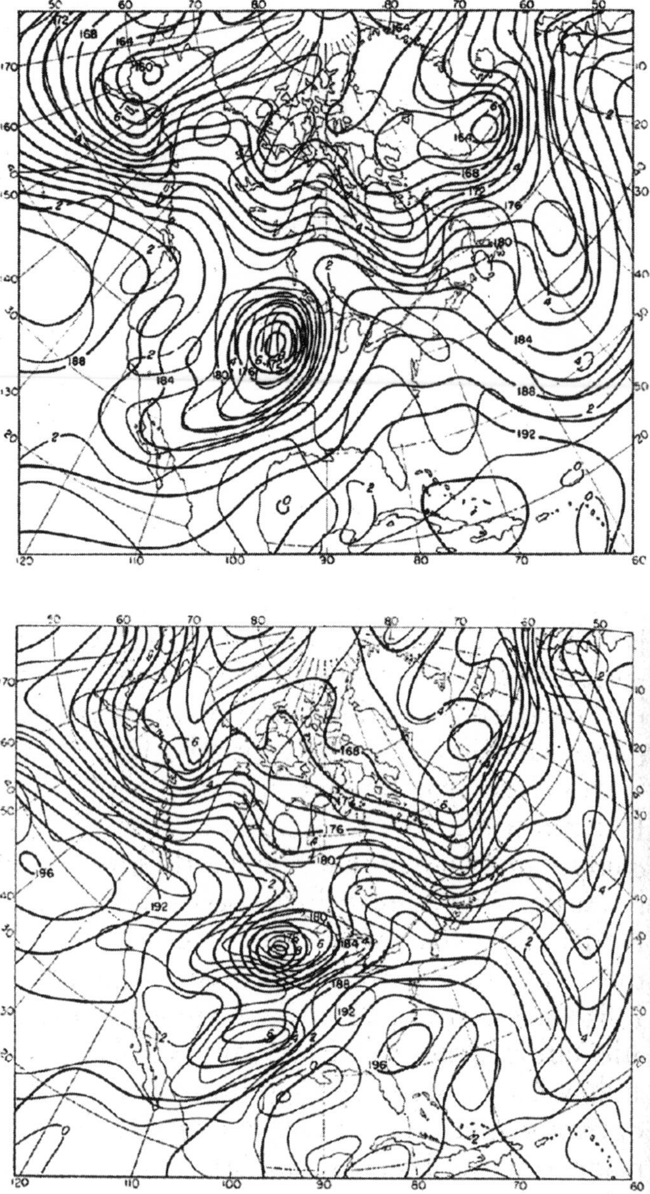

Figure 6.16. The "acid test": The top chart is the analysis of the synoptic situation at 0300 hrs on January 5, 1949, and the bottom chart is the forecast for this time as computed by ENIAC. The thick lines are the 500 mb geopotential height, and the thin lines are total vorticity. The charts are reprinted from J. G. Charney, R. Fjørtoft, and J. von Neumann, "Numerical Integration of the Barotropic Vorticity Equation," *Tellus* 2, no. 4 (1950): 237–54. Reprinted with permission.

amplification of a trough, a valley of low pressure extending for many hundreds of kilometers over the western United States at about 110 degrees west, was well predicted, as were the large-scale shifting of the wind from NW to WSW and the increase in pressure over western Canada. Over the North Sea on that day, the amplification of a trough was correctly predicted, along with the breakdown of the tip of a high-pressure system over France. The next day the forecast was even better; the continued turning of the northwesterly winds over the North Sea and their extension into southwestern Europe was correctly predicted. Successful prediction of pressure troughs over land usually led to better rain or snowfall predictions, so this was very encouraging.

On February 13 the main forecast and observed changes occurred over the west coast of North America and in the Atlantic Ocean and were consequently difficult to forecast and verify. Of course, this problem was a major one for operational considerations because it highlighted the difficulties introduced by the lack of data over the oceanic regions where much of the weather for North America and Europe originates.

The first powered aircraft flights had lasted only hundreds of meters and often ended with crashes. The first electronic calculation of weather had similar limitations, but it had worked! Now more effective "forecasting machines" could be developed, and the age of computer forecasting had dawned. Within a couple of years, numerical forecasts were being produced by several groups in Europe and America. Machines rapidly increased in power and capacity, and, hence, the capability of the simulations to predict more detail also increased.

In the paper published in *Tellus*, Charney and his colleagues discussed the shortcomings of the forecasts, and they tried to account for the errors that had occurred. Reminiscent of Richardson, they identified (1) the shortcomings of the model in that it had omitted key physics and dynamics; (2) the errors inherent in the initial charts from which the forecasts were made; (3) the errors in the conditions imposed at the boundaries of the domain; and (4) the errors associated with replacing differential equations by finite-difference equations.

These issues have turned out to still have relevance today. Let us deal with the last category first. Charney and colleagues referred to the differences that arise from replacing differential equations by finite difference equations as *truncation errors*. These "errors" are the inevitable consequence of using weather pixels (of finite size) to represent smoothly changing variables. The calculation evolves the pixels rather than the original variables; consequently, there is the potential for the finite differences to distort the final result. One of the ways they went about trying to ascertain the impact of this on their forecasts was to examine the values of the vorticity field—the point being that the exact differential equations express a conservation law for total vorticity, therefore the values of the total vorticity should be conserved as they move with the flow. Upon examination of the charts, Charney's team detected changes to the minimum values of the vorticity that were present in the initial charts. These should have been conserved during the forecast, but they were not. They also noted changes in the vorticity pattern for the forecast of the depression on January 5 that had to be due to truncation errors.

In general, it is very difficult to attribute parts of an error to each particular source that Charney and his colleagues identified. An important factor that influenced their considerations was the simplicity of the model itself, factor 1. That is, they had used Rossby's conservation law, which assumes that there is no variation in wind with height and no mechanism for representing the effects of heating and cooling on the wind. In the real atmosphere, total vorticity is not strictly conserved and there exist mechanisms, such as moisture processes, for creating and destroying vorticity. Ignoring these effects in a model implies that the model simulation must depart from reality at some stage. Before the experiments were carried out, it was felt that because they were focusing only on the largest-scale features in the atmosphere, and because they were only trying to simulate the behavior of one variable (the vorticity) at a particular height, Rossby's equation would then provide a suitable model for their purposes. However, they concluded that sometimes the effects of heat and moisture processes could not be ignored over time scales of twenty-four hours in the midtroposphere. Even when these heat and moisture effects are small, they may be responsible for important structural changes in the wind field—the small vertical motions that were ignored earlier in figure 6.12 actually sometimes matter a lot.

A major source of uncertainty in the forecast is the specification of the initial conditions, factor 2. If the chart from which a forecast is started is in error, there is little hope of the forecast being completely correct. For example, a low-pressure system moving over North America may move into the Atlantic, where it is less well observed (this was certainly true in 1950, and is still partially true today because satellites find it difficult to probe all levels of the atmosphere). The ENIAC calculations were initiated from the hand-drawn forecasters' charts, and the values of the variables indicated by contours on those charts had to be interpolated to the grid points of the model. There is considerable scope for introducing errors in this process, which amount to not only getting the grid point values "wrong" but also—and in many ways this is more serious—getting the values from one grid point to the next inconsistent with the observed weather pattern.

This second point is critical. We have to remember that the atmosphere exhibits large-scale balances, as we discussed earlier in this chapter, so that the averaged wind field, for example, is usually close to geostrophic balance. This implies that there is naturally a constraint on how the values of the height fields can change from one pixel to the next. Where observations are plentiful, the forecaster can usually estimate fairly accurately what the value of a particular variable at the nearest pixel should be. But where the observations are sparse, or when dealing with pixels a long way from a high density of observations, the problem becomes much more acute. In modern operational forecasts, observations are given differing importance according to their distance from a pixel, and according to how much faith we have in them, as we discuss further in chapter 8. But, back in 1950, they had to rely on manual methods of interpolation—mostly educated guesswork. In the remainder of the book we describe how the emphasis moved from how to get the "right" forecast to how to get the "best" forecast.

Charney sent a copy of the results to Richardson, who lived just long enough to see the beginnings of the fulfillment of his "mere dream." Richardson asked his wife, Dorothy, to comment on the results from Charney's team. She concluded that the computer model had done a somewhat better job when compared with charts that would have essentially led to a forecast of persistence—that is, the weather tomorrow is the same as the weather today. Richardson wrote back to Charney, in one of his last scientific letters, congratulating him on the success and added graciously in conclusion that the ENIAC calculations were a huge advance on his own work, which had foundered at the final hurdle.

Subsequent progress with this early numerical forecasting was very encouraging, although trials continued with comparison to known observations for some time. A particularly severe snowstorm over Washington, D.C., occurred on November 6, 1952, and reproducing this storm a year later proved to be a stringent test for the new models. Once again, Charney was involved, but he had obviously not lost any of his excitement about the whole enterprise. In the early hours one morning, the charts were completed for the 1952 storm, and Charney phoned Harry Wexler, the director of research for the U.S. Weather Bureau, exclaiming "Harry, it's snowing like hell in Washington." Harry, who had been enjoying a good night's rest, was presumably delighted to hear this.

On April 28, 1954, forty years after the publication of Vilhelm Bjerknes's paper entitled "Meteorology as an Exact Science," the president of the Royal Meteorological Society in London, Sir Graham Sutton, gave the annual presidential address to the society, titled "The Development of Meteorology as an Exact Science." He defined an "exact science" as one admitting quantitative treatment, and he distinguished it from a purely descriptive summary of knowledge gleaned from observations. He identified exact sciences as branches of mathematical physics and lost no time in questioning the extent to which meteorology (and forecasting in particular) had attained such a status. A transcript of his address was published in the *Quarterly Journal of the Royal Meteorological Society* later that year. In the lecture, Sutton refers to "L. F. Richardson's strange but stimulating book"—referring, of course, to the 1922 publication *Weather Prediction by Numerical Process*—and goes on to remark that "I think that today few meteorologists would admit to a belief in the possibility (let alone the likelihood) of Richardson's dream being realised. My own view, for what it is worth, is definitely against it." Strong sentiments, and they were something of a "shock amid the new" because Sutton was also the director-general of the British Meteorological Office. At the dawn of the modern era of computing in the early 1950s, the Met Office was embarking on research into numerical weather prediction. British scientists were joining many others around the world in an effort to finally vanquish the capriciousness of weather and its chancy forecasting. Armed with the new computers and half a century's worth of meteorological theory born out of Bjerknes's program, it was looking to many as though weather's number was finally up.

So, following the modest success of ENIAC, why was a senior civil servant such as Sutton being so pessimistic about a promising new development in an area of science that would benefit us all? He certainly acknowledged the potential of applying computers to solving the problem of numerical weather prediction, and he described some of the first work to be carried out on this problem by his staff at the Meteorological Office. But even with some of his best scientists working on these problems, Sutton's final prognosis for an upturn in the fortunes of the weather forecasters was distinctly dull and gloomy. Sutton had taken account of the technological advances: his pessimism was engendered by a pragmatist's vision of the feasibility of numerical forecasting. He emphasized the importance of instabilities in the atmosphere as described in the theories of Charney and Eady—the telltale signs of which were subtle patterns in the wind and temperature fields—and then came to the logical conclusion that the paucity of observations made the prospect of actually forecasting the growth and development of weather systems highly unlikely. And this was an entirely reasonable conclusion. Satellite-borne instruments now help to redress this lack of data. Further, modern improvements to radar and the development of sensors that scan the atmosphere at different wavelengths are helping us see the atmosphere below the tops of clouds, which is where much of the interesting weather develops.

Nevertheless, in 1950 a landmark had been created by the Princeton team in the history of theoretical meteorology and practical weather forecasting. Quite surprisingly, it had been achieved with a very simple yet mathematically robust model of the atmosphere. Useful information had been extracted from what was otherwise a very oversimplified mathematical view of "weather." Perhaps what is more surprising is the attitude that prevailed at the time concerning the usefulness of the vorticity conservation law. It was assumed that it would more or less only perform with the accuracy of linear prediction. The subtleties of the nonlinear feedback mechanism between the vorticity and the wind field that moves the vorticity were not anticipated.

The initial results from ENIAC, although encouraging, had taken almost as much time to compute as the weather took to change, and the end result was certainly no better than a human forecaster could have achieved. Although it took so long for the ENIAC forecast, much of the time was consumed by manual punch-card operations. Namely, the reading, printing, reproducing, sorting, and interfiling operations of the cards were all done by hand, and about one hundred thousand cards were used in one twenty-four-hour forecast. Even by 1950, Charney and his colleagues realized that the type of machine that was being built for the Institute for Advanced Study would not require this manual secretarial work and would reduce the time by half—a twenty-four-hour forecast might be produced in only twelve hours and would therefore be an actual forecast. (We mention that Peter Lynch, of Met Èireann and University College, Dublin, put the original ENIAC calculation on a mobile phone in 2008. It is able to reproduce the ENIAC results, including a picture, in just a few seconds! He calls this PHONIAC, and the process is nearly one hundred thousand times faster than ENIAC managed.)

These very simple models ignored one basic and important physical process, as Charney had shown. The northward variability of temperature allowed thermal potential energy to transfer to the wind field as movement, or kinetic, energy. This transfer of energy is crucial in the development of storms and hurricanes, and the development of cyclones in general. So computer prediction of cyclone development was severely restricted until an effective method was found to "layer" the atmosphere in the vertical, such as that described near the end of the previous chapter.

Before numerical weather prediction could become a widespread, established practice, computer models had to advance to the stage where multiple height or pressure levels in the atmosphere were feasible, thereby providing the forecaster with a major weapon—the surface pressure chart. The crucial step was to predict the motion not only of a pressure surface about halfway up the troposphere but also that of a pressure surface nearer the ground. The two-level model, as it became known for obvious reasons, connected what was happening midway up the atmosphere with what was happening at the Earth's surface. The key variable that provides the link is vertical motion: if there were no vertical motion, then the motions at different levels would be independent of one another, and we know from observation that this is not true. The surprising solution to this difficulty was to abandon the attempt to directly compute vertical motion except by indirect adjustment to computed convergence levels in each layer. A three-layer model, used extensively by the Swedish researchers, proved remarkable effective. As computers increased in power and researchers became more confident, more and more layers were built into models. By the 1970s, typically ten to twelve layers were considered, and by the beginning of the twenty-first century, most weather centers were using more than fifty layers.

So now we have the equations, and we have the technology, especially the satellites and computers; but we always have unknowns and errors, and we strive continually to improve the physics. In his 1951 paper "The Quantitative Theory of Cyclone Development," Eady, with characteristic depth of insight, observes:

> The practical significance of a demonstration that the motion is unstable is clear, for in practice, however good our network of observations may be, the initial state of motion is never given precisely and we never know what perturbations exist below a certain margin of error. Since the perturbation may grow at an exponential rate, the margin of error in the forecast (final) state will grow exponentially as the period of the forecast is increased.... After a limited time interval ... the possible error will become so large as to make the forecast valueless.

He goes on to conclude the paragraph by noting, "Thus long-range forecasting is necessarily a branch of statistical physics in its broadest sense: both our questions and answers must be expressed in terms of probabilities."

Eady was alluding to the effects of chaos. These issues were taken up in earnest within a few years by a then relative unknown, Edward Lorenz at MIT, as we discuss in the following chapter.

Figure CI.1 The atmospheres of Earth and Jupiter are conspicuously different, but they are governed by the same basic physical laws. Using math, we are able to isolate the distinguishing features that lead to the radically different patterns we observe. By accounting for the rotation of the planets, the depth of their atmospheres, and the role of vortices, many of the salient differences can be quantified. Photos courtesy of NASA.

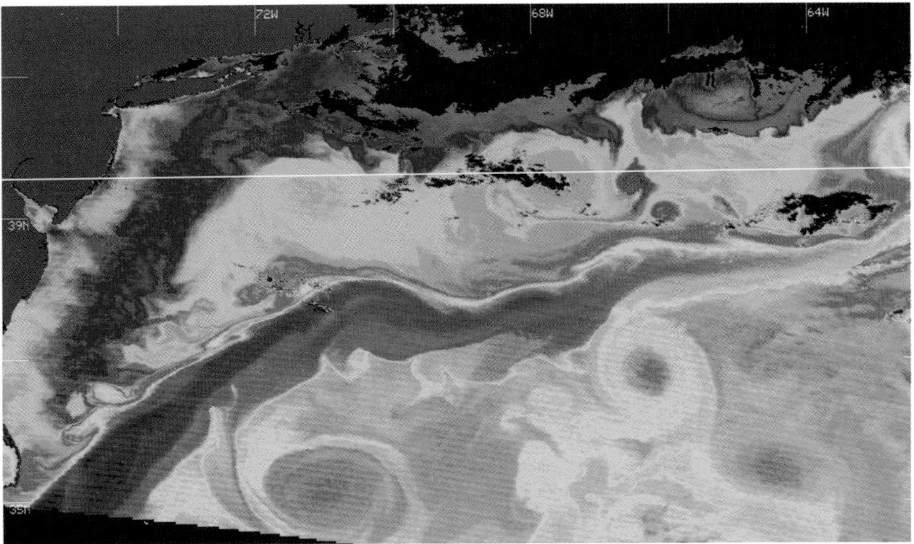

Figure CI.2. Swirling eddies in the Gulf Stream off the East Coast of the United States. The red color shows the warmer water of the Caribbean being carried by the Gulf Stream across the northern Atlantic Ocean. The density of ocean water is also dependent on its salinity—here the blue shows colder and fresher water. The eddies are analogous to the low-pressure systems seen in the atmosphere in figure 1.3. The sea surface temperature image was created at the University of Miami using the 11- and 12-micron bands, by Bob Evans, Peter Minnett, and co-workers. Reprinted courtesy of NASA.

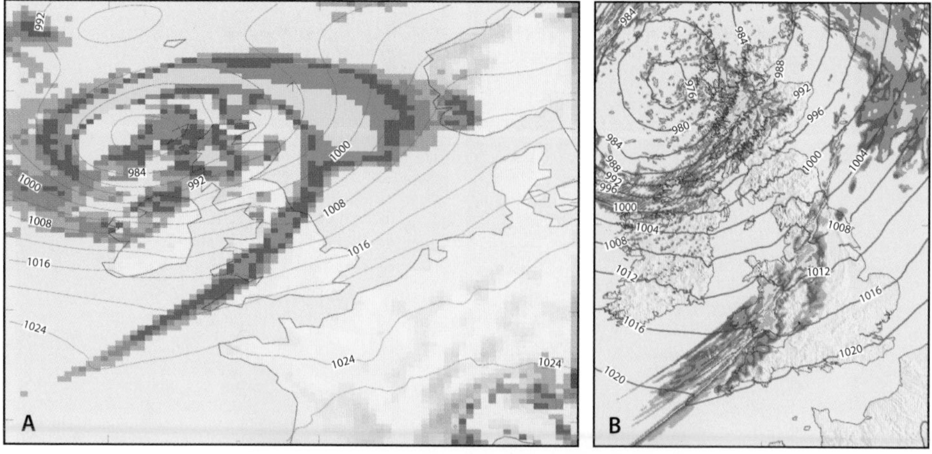

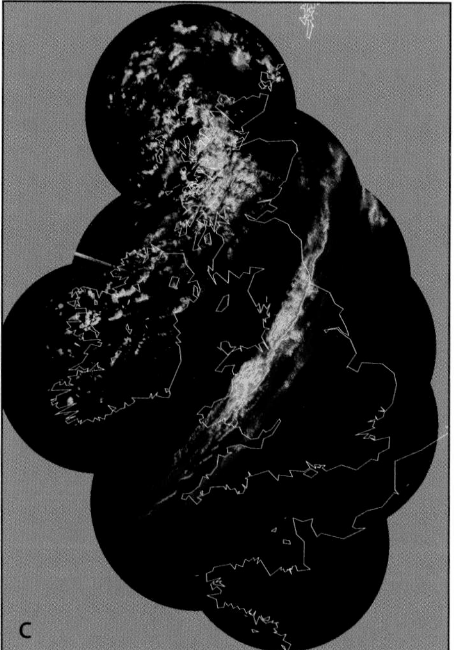

Figure CI.3 a, b, and c. These three figures show two forecasts of a band of rain over eastern England (figures a and b) and the radar image (figure c) showing the observed precipitation on May 23, 2011. The yellow/red shades indicate heavier rainfall. The contours on figures a and b are isobars of mean sea-level pressure. The mosaic patterns reflect the finite resolution of the models, predicting the rainfall at the scale of the weather pixels. A global model with a horizontal resolution of 25 kms produced figure a, and a regional model with a horizontal resolution of 1.5 kms produced figure b. Observe the improvement between them when they are compared with the "reality" shown in figure c; the better forecast in figure b has come from the increased resolution of the weather pixels.

Figure CI.4. Richardson's vision of a "forecast factory" in a place such as the Royal Albert Hall in London. Sixty-four thousand workers are arranged in tiers of five circles. Each group of people is responsible for calculating the weather for their part of the globe, and then passing this information on to their neighbors. The "conductor" is orchestrating the huge number of calculations so that each group performs calculations at the same time, and no group gets out of step. In modern parallel computer processing setups, the workers are replaced by individual computers. *Le guide des Cités* de François Schuiten et Benoît Peeters © Casterman. Avec l'aimable autorisation des auteurs et des Editions Casterman.

Figure CI.5. This image of the Earth's atmosphere was taken by a Japanese weather satellite and shows the eruption of Mount Redoubt in Alaska on March 26, 2009. The volcanic ash plume rises to about 65,000 feet (19.8 km), which is a considerable proportion of the depth of the atmosphere visible from the satellite. The troposphere, the lowermost portion of the atmosphere in which most weather as we recognize it occurs, is a mere 9 km deep. The Earth, with an average radius of 6,341 km, is therefore surrounded by a very thin "shell" of fluid. This relative thinness of the atmosphere, combined with the fact that it surrounds a planet that spins reasonably rapidly, explains why hydrostatic and geostrophic balances are so important in the basic equations. Photo courtesy of NASA.

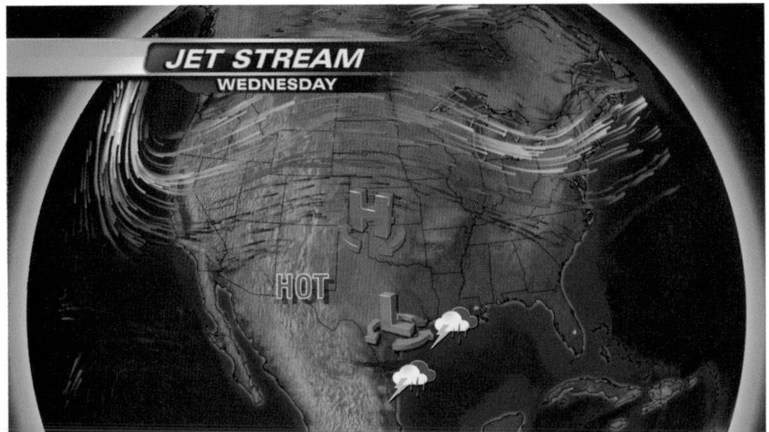

Figure CI.6. An idealized weather graphic shows high and low pressure systems over the panhandle and southern Texas. The jet stream, centered in the lighter blue streamlines, exhibits a pattern characteristic of a Rossby wave.

Figure CI.7. The large-scale swirling cloud structure over the British Isles imaged by a satellite is of paramount importance to the forecaster. Much smaller patterns in the clouds can also be seen; although they may influence local conditions, they are of secondary importance to the forecaster interested in predicting the weather for the next day or two. The continuous band of cloud is associated with the warm front, while cooler, showery conditions follow behind from the northwest. © NEODAAS / University of Dundee.

Figure CI.8. When viewed from space, rippled clouds (in the center of the picture) at a height of 2 kms over the Scottish mountains reveal gravity waves. Gravity waves do not transport air horizontally, nor do they significantly mix different air masses. Their role is to help the atmosphere adjust to the planet's motion, but that is another story. © NEODAAS / University of Dundee.

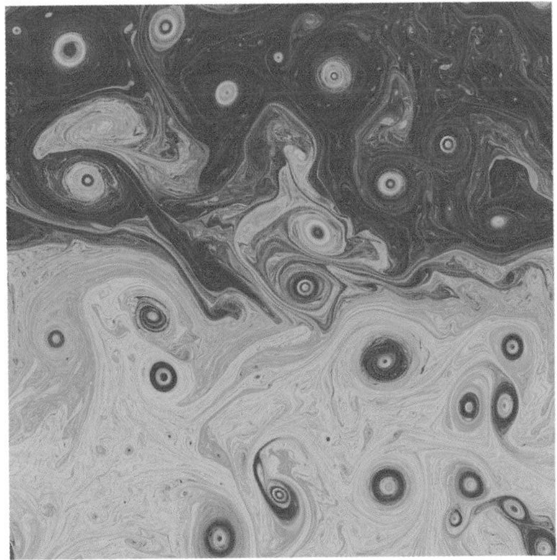

Figure CI.9. The background (blue above / yellow below) shows a simple left to right (sheared) flow. A modern computer calculation needs to capture, preserve, and follow all the tiny superimposed vortex swirls in order to represent the turbulence generated in the flow. The intricate detail here is simulated using an area-preserving computational method. © David G. Dritschel.

Figure CI.10. The interior of the nave of Lavenham Church, Suffolk, England, constructed in the "perpendicular style" of English church architecture. Simple geometric principles—the emphasis on vertical mullions and horizontal transoms in the windows—define the style. Here the geometry helps us to focus on the principal static forces that hold the weight of the building up against gravity. Once the geometric principles are incorporated into the basic design, the stone masons can unleash their creative skills and add the fine artistic detail. © Claire F. Roulstone.

Figure CI.11. The Golden Gate Bridge, San Francisco. Modeling such cloud layers uses simple geometric ideas based on scaling for the buoyancy forces. While the fine detail of the cloud pattern changes from day to day, it does so in a way that respects geometrical rules and the physics of water vapor. Photo by Basil D Soufi.

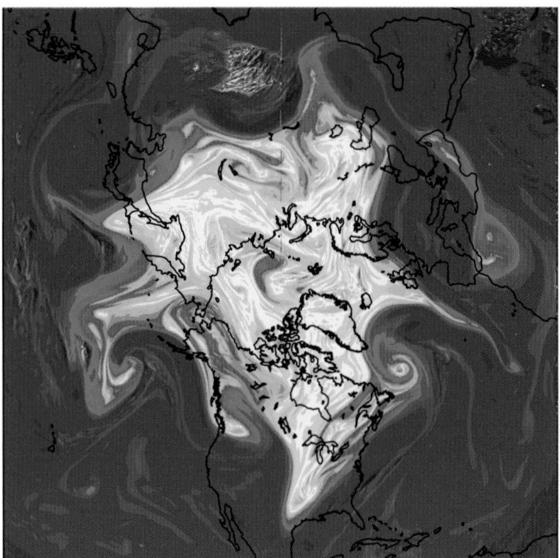

Figure CI.12. PV (on a 315K potential temperature surface) is obtained by processing the wind and temperature fields from a numerical forecast on February 6, 2011. It is illustrated over a view of Earth taken from above the North Pole. Transforming from the basic wind and temperature variables to PV enables forecasters to assess how well a numerical model is capturing the major features of cyclone development, which show here as swirls on the edge of the yellow-green colors. © ECMWF. Reprinted with permission.

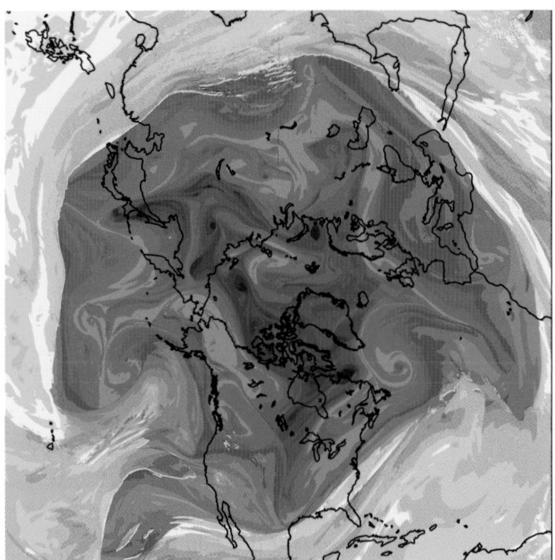

Figure CI.13. A plot of potential temperature (colored) on a surface of constant potential vorticity (effectively, the tropopause where PV = 2). These maps are an invaluable aid to the forecaster, especially in predicting the location and behavior of jet streams on the yellow/green and dark blue boundaries. Such plots show how dynamic and thermodynamic variables can be combined to give a "top-down" view of the atmosphere; that is, they show the mechanisms responsible for the development of major weather systems. © ECMWF. Reprinted with permission.

Figure CI.14. Heavy rain moving across the east Aegean, viewed from Chios harbor, looking toward the Turkish coast on October 28, 2010. Sailors and mountaineers are particularly vulnerable to sudden changes in the weather; the challenge for modern computer simulations is to predict where and when these changes will occur. The band of heavy rain shown here has a sharp leading edge, and tracking this feature is a challenge for weather prediction models. © Rupert Holmes.

Figure CI.15. From solar system to Earth system—our story describes how mathematicians are now focusing on ways to represent the detailed interactions and feedbacks that control our environment. According to Greek cosmogony, Gaia (the Earth) sprang from Chaos (the great void). Modern science explains how heat and moisture processes create the clouds and power the winds; the swirls in the clouds are created by the wind, which redistributes heat and moisture around the planet. What sort of unpredictability do these interactions lead to? Math helps us decide what we can say about the future of our planet. Water image courtesy of Monroe County, Tennessee, Department of Tourism; life image © Michael G. Devereux; earth image courtesy of NASA; land image courtesy of the U.S. National Park Service; air image © Rupert Holmes.

SEVEN

Math Gets the Picture

Charney's 1948 paper "On the Scales of Atmospheric Motion" reveals how hydrostatic and geostrophic balance, together with the conservation of potential temperature and potential vorticity, set us on the road to understanding much about temperate latitude weather. Fifteen years later, Edward Lorenz published a paper under the somewhat innocuous title "Deterministic Nonperiodic Flow," and concluded that long-range weather forecasting might be forever beyond our capabilities. Would chaos undo all that had been achieved in theoretical meteorology? In this penultimate chapter, we show how the rather different results and conclusions of Charney and Lorenz can be rationalized within a single mathematical framework. Using mathematics we are able to capture the qualitative features of weather systems, together with their regular and irregular behavior, in a way that enables us to build this into computer simulations.

A Butterfly Emerges

Around 700 BC the Greek philosopher Hesiod wrote *Theogeny*, an account of cosmology and cosmogony, in which he offered an explanation as to how the gods came about. He introduced the entity "chaos," a yawning gap between heaven and Earth, which was nothing but "formless matter and infinite space." If heaven was envisioned as a place where uncertainty was vanquished, and if the harsh realities of life on Earth meant that tomorrow was not certain, then chaos was the obstacle that denied mortals a deterministic view of the universe. Chaos is popularly

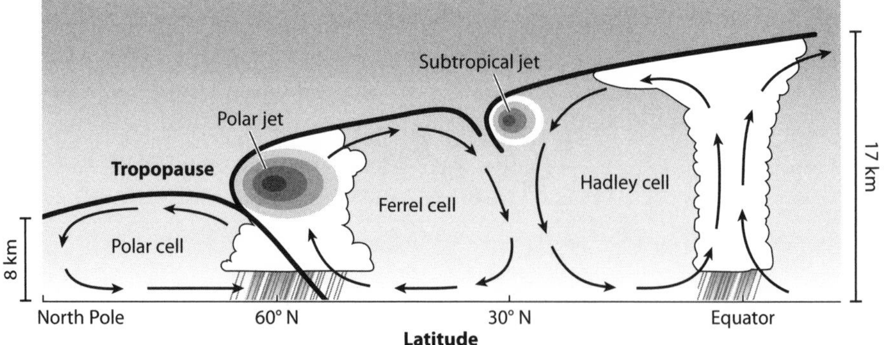

Figure 7.1. This idealized sketch represents an averaged April or May behavior in a cross section of Earth's atmosphere extending from the North Pole to the equator. Convective clouds are created by updrafts of warmer moist air in the tropics and midlatitudes. Lorenz's model of convection is a simple model of one of these "cells." Note how the height of the clouds varies from equator to pole because of the differential heating. Scaling rules quantify this variation of height, allowing us to make such universal sketches in spite of convection currents that may be chaotic.

seen as standing in the way of science and our ability to comprehend the world around us. One of the reasons why chaos has attracted so much attention in recent years is that it is often seen as pulling the proverbial rug from under the feet of scientists: it affirms what we all "know"—that the science of prediction in areas such as economics, medicine, and weather forecasting is by no means infallible. Yet chaos, and the mathematics we use to study it, is anything but formless. Within the apparent void lie structure and a rich tapestry of very practical mathematical ideas.

Charney and Eady had appreciated that unavoidable inaccuracies would creep into the weather forecasts. The lack of a complete understanding of the physical processes involving clouds, rain, and ice as well as of the wind gust interactions with the trees and hills, for instance, together with rational approximations resulting from the computer-based methods always limits the range over which useful weather forecasts can be made. The chaos story started in 1972, when the American meteorologist Edward Lorenz remarked that the flapping of a butterfly's wings over Brazil might trigger a tornado over Texas. (In fact, it was a conference organizer who introduced the "butterfly wings"; Lorenz had failed to submit a title and abstract for his lecture in time for the publication of the program, and so the organizer, who knew that Lorenz had used a seagull's wings as an analogy in his previous talks, improvised.)

Back in the 1950s, Lorenz had been looking for periodic patterns in the weather, just as Poincaré had been looking for periodic solutions to the three-body problem. Lorenz's original goal was to show that numerical models based on the predictive laws of physics would always prove to be superior to statistical methods of forecasting based on information about past events. Consider cloudy days in Seattle. If on average it is cloudy for ten days during July, could a computer solution of the weather equations give us better odds on predicting cloud cover there for a given week?

To test this conjecture, Lorenz decided to search, using a numerical model, for a *nonperiodic* solution of the equations governing a very simple model of weather. A nonperiodic solution is one that never quite repeats itself. The crucial point was that if the atmosphere exhibited nonperiodicity, then no matter how good prediction methods using past records might be, they could never capture nonperiodic behavior. This is because, by definition, a future state of the system would never be quite the same as any previous state from which the data was obtained.

By the late 1950s, with computer power attaining useful levels, Lorenz had the tools at his disposal for such a task. He had access to a computer capable of sixty multiplications per second and a memory of four thousand words—primitive by today's standards but sufficient to create a model of a very simple "atmosphere." The model that Lorenz chose is one in which the only form of "weather" is the overturning of a fluid when

Figure 7.2. Edward N. Lorenz (1917–2008) was born at West Hartford, Connecticut, and received his higher education at Dartmouth College, Harvard University, and MIT. Lorenz was a member of the staff of what was then MIT's Department of Meteorology from 1948 to 1955, when he was appointed as an assistant professor. He received numerous awards and prizes, and in 1991 he was awarded the Kyoto prize. The committee remarked, "Lorenz made his boldest scientific achievement in discovering 'deterministic chaos,' a principle which has profoundly influenced a wide range of basic sciences and brought about one of the most dramatic changes in mankind's view of nature since Sir Isaac Newton." Photo courtesy of the MIT Museum.

heated from below. This type of motion occurs when we gently heat a pan of soup. In the atmosphere, hot air rises and cool air descends. This happens at many places every day, giving rise to cumulus clouds and sometimes rain or snow, or even occasionally the violent localized phenomena of thunderstorms and tornadoes.

Lorenz's convection currents were imagined to be roughly circular and much more regular, as indicated by the cells in figure 7.1: cold air from the top of the atmosphere was descending in one region and warm air from near the ground was rising in the other region. His model atmosphere was contained by two horizontal surfaces that were heated differently—just like the liquid in a saucepan where the bottom of the pan is being heated, and the top of the liquid is cooling. Lorenz decided to simulate the essential characteristics of the motion of the fluid by reducing the mathematics to just three equations for the variables X, Y, and Z: X measures the intensity of the convective motion; Y measures the temperature difference between the ascending and descending currents; and Z measures the amount by which the variation of temperature with height departs from a linear relationship.

The three Lorenz equations, given in figure 7.3, couple the rates of change dX/dt, dY/dt and dZ/dt to the variables X, Y, and Z through deceptively simple algebraic expressions of addition and multiplication. But they turn out to have a complicated and rich mathematical structure, which has been the subject of thousands of scientific papers since

Lorenz Equations

$$\frac{dX(t)}{dt} = \sigma Y(t) - \sigma X(t)$$

$$\frac{dY(t)}{dt} = \rho X(t) - Y(t) - X(t)Z(t)$$

$$\frac{dZ(t)}{dt} = X(t)Y(t) - \beta Z(t)$$

Figure 7.3. The three variables X, Y, and Z all depend on time t. $X(t)$ represents the intensity of the convection currents, $Y(t)$ represents the temperature difference between the ascending and descending currents, and $Z(t)$ represents the difference between the actual fluid temperature and a fixed linear temperature profile. The three constants σ, ρ, and β determine the behavior of the solution. The two product terms, $-X(t)Z(t)$ in the second equation and $X(t)Y(t)$ in the third equation, represent the nonlinear feedback in the system. Computer solutions of the Lorenz equations produce values of $X(t)$, $Y(t)$ and $Z(t)$—the lists of numbers at specified intervals, or values of time, t, given in table 1 of figure 7.4. Only one out of every five numbers was actually listed by Lorenz to save space.

1963. And that richness comes from the nonlinearity, which arises from the terms $X(t)Z(t)$ and $X(t)Y(t)$ shown in figure 7.3. One of the feedback processes involves the response of the internal temperature when the upwelling increases. As the hotter fluid is carried up more quickly, this modifies the very temperature profiles, and the difference, that caused the upwelling. This changes the feedbacks, and so on. But what are the consequences for forecasting future states?

The numerical procedure Lorenz adopted advanced the "weather" in six hourly increments, or time steps, and he programmed the computer to print out three variables every fifth time step, or once per simulated thirty hours. Simulating this graphic "weather" for one day took about one minute of computer time, and printing the numbers was actually much more time-consuming than the arithmetic. To make the printout of the results easy to read, he wrote numbers to three decimal places; for example, 1.078634 was printed as 1.078.

After accumulating many pages of output, Lorenz wrote another program to plot the results as graphs to make it easier to spot patterns in the numbers (see figures 7.4 and 7.5). Lorenz examined these emerging patterns, and, indeed, they appeared to be random, just as we might expect the overturning eddies in a gently boiling kettle not to recur precisely. One day he decided to take a closer look at a set of results because he wanted to make sure his experiments were reproducible. He repeated an experiment by taking the numbers that came out of the machine and putting them back in as starting conditions for a new run. Lorenz then expected to see the same curve being traced out a second time. At first the solutions were the same, but then they began to move apart until there was no resemblance between them at all—just as happened in the double pendulum experiment described near figure 4.6 (see also figure 7.14).

Was this an accident? He tried again and got the same result. Lorenz suspected that he might have introduced small errors when he input the data, but he would not have believed that this could lead to wildly differing answers. It had been generally accepted since the time of Newton that small errors would only cause small effects. In this case, though, the effect was huge. He then suspected a technical problem with the computer. Perhaps one of the valves was overheating, which was not uncommon.

Before calling in an engineer to examine the computer, Lorenz set about finding how else such an error might occur. After some thought,

TABLE 1. Numerical solution of the convection equations. Values of X, Y, Z are given at every fifth iteration N, for the first 160 iterations.

N	X	Y	Z
0000	0000	0010	0000
0005	0004	0012	0000
0010	0009	0020	0000
0015	0016	0036	0002
0020	0030	0066	0007
0025	0054	0115	0024
0030	0093	0192	0074
0035	0150	0268	0201
0040	0195	0234	0397
0045	0174	0055	0483
0050	0097	−0067	0415
0055	0025	−0093	0340
0060	−0020	−0089	0298
0065	−0046	−0084	0275
0070	−0061	−0083	0262
0075	−0070	−0086	0256
0080	−0077	−0091	0255
0085	−0084	−0095	0258
0090	−0089	−0098	0266
0095	−0093	−0098	0275
0100	−0094	−0093	0283
0105	−0092	−0086	0297
0110	−0088	−0079	0286
0115	−0083	−0073	0281
0120	−0078	−0070	0273
0125	−0075	−0071	0264
0130	−0074	−0075	0257
0135	−0076	−0080	0252
0140	−0079	−0087	0251
0145	−0083	−0093	0254
0150	−0088	−0098	0262
0155	−0092	−0099	0271
0160	−0094	−0096	0281

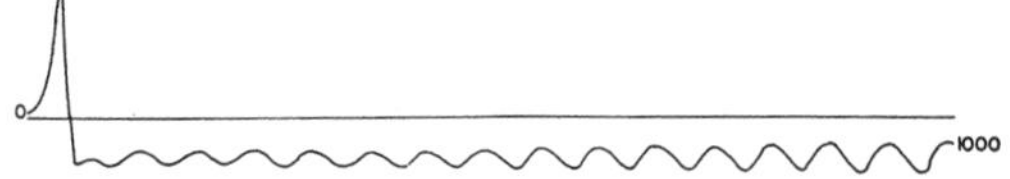

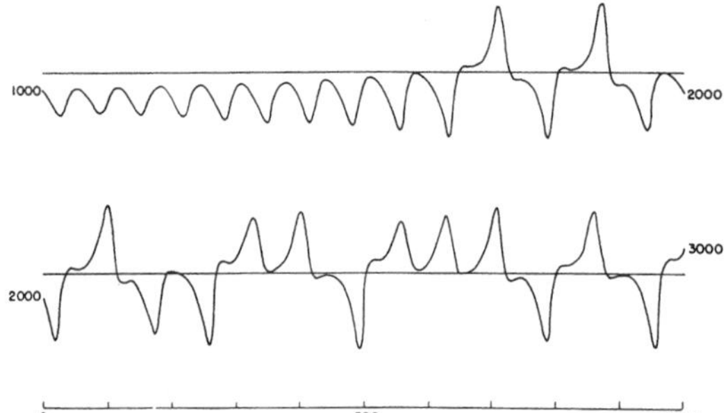

FIG. 1. Numerical solution of the convection equations. Graph of Y as a function of time for the first 1000 iterations (upper curve), second 1000 iterations (middle curve), and third 1000 iterations (lower curve).

Figure 7.4. Copies of the actual results from Lorenz's 1963 paper. The initial rise and then fall of the values for Y given in the third column of table 1 are plotted in his figure 1. The remainder of the graph depicted in figure 1 shows the apparent random behavior of this variable: no clear pattern is discernible from a study of this graph alone. In fact, Lorenz and some of his colleagues challenged each other to predict when the curve would next switch signs. No one found a winning strategy. *Journal of the Atmospheric Sciences* 20 (1963): 130–41. © American Meteorological Society. Reprinted with permission.

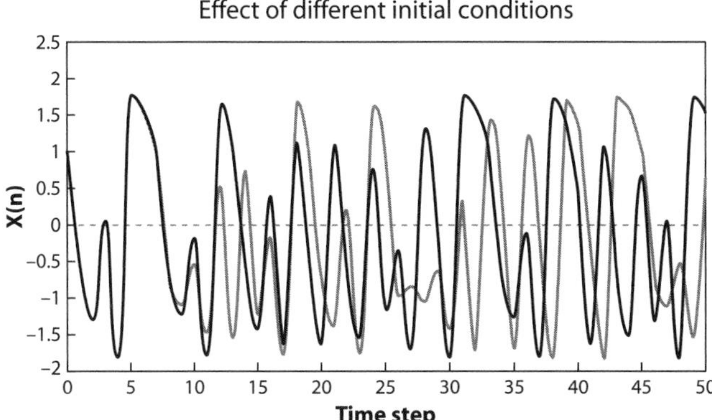

Figure 7.5. Two solutions of the Lorenz equation (one given by the black curve and the other by the grey curve) that start from two slightly different initial conditions. At first the two plots overlap, but gradually they diverge around time step = 10, showing the effect of the very small initial difference. By the end, there is no apparent relation between the curves. In fact, both curves appear quite random in their roughly oscillatory behavior. © ECMWF. Reprinted with permission.

he realized that the calculations being performed within his computer were using six decimal places, but he had taken the numbers with three decimal places and used these as a starting point for checking the earlier calculations. Lorenz considered the possibility that the error introduced by using only the first three decimal places, an error of about 0.2 percent, might be responsible for the breakdown of his "identical twin" experiments.

It turned out that these very small errors had profound consequences, and they *were* responsible for the divergence of the solutions. Lorenz had discovered sensitive dependence of solutions on the starting states, and thus chaos. But surely we can improve the accuracy of our calculations by including the missing digits? Lorenz spotted the crucial implication for weather forecasting: while an error of one part in a thousand might have been correctable in his toy model, much bigger errors are unavoidable in real-world forecasting. We can never measure weather pixel values with 100 percent accuracy, and there are far fewer observations than data grid points in the models—often data is missed over inhospitable land and sea surfaces. In the conclusions to his 1963 paper, Lorenz says, "When our results concerning the instability of nonperiodic flow are applied to the atmosphere, which is ostensibly nonperiodic, they indicate that prediction of the sufficiently distant future is impossible by any method, unless the initial conditions are known exactly. In view of the inevitable inaccuracy and incompleteness of weather observations, precise very-long-range forecasting would seem to be non-existent." So, did his discovery uncover the great chasm of chaos between deterministic science and reality? Were Bjerknes's and Richardson's dreams for forecasting the weather dashed and in ruins?

A commonly accepted view is that this does spell major trouble for present-day forecasters. But Lorenz's 1963 paper is not famous just because of its earth-shattering conclusions concerning predictability. Indeed, the most important feature of Lorenz's 1963 paper from our recent perspective is the way he went about studying his toy model of weather. If we take a closer look at both the nature of Lorenz's results and how he interpreted them, we discover that he developed novel methods for extracting information from systems that are liable to chaotic behavior.

So how did Lorenz apply Poincaré's ideas? Lorenz's search for nonperiodic solutions to the equations was facilitated by the new way that he viewed the computer output. In figure 7.5, the solutions were plotted against time and showed apparently formless but very roughly oscillating behavior. In figure 7.6, the results of his numerical integrations were plotted as graphs where the axes are chosen to represent two of the variables, and the curve evolves with time as each new point—$Y(t)$, $Z(t)$—is plotted. In this way, the solutions appear as curves suspended in the space spanned by these axes. This representation would help him to spot when solutions had regular patterns because each would overlay itself if its behavior repeated. However, the curves never quite overlapped as time progressed, and Lorenz had found nonperiodic behavior for the convecting cell.

More importantly, Lorenz discovered that the apparently random curves, or life histories, of figures 7.4 and 7.5 actually had a new pattern. Most of the oscillations, when plotted in the three-dimensional space made of the X, Y, and Z variables, lay near to the "butterfly's wings" of figure 7.7. The chaotic behavior showed up in terms of the *transition* from one wing to the other. In Lorenz's experiment, this transition means that the sense of rotation of the fluid cell changes. His remarkable picture captures an important qualitative feature of the physical system that was not apparent in the traditional methods of displaying output. Lorenz applied the abstract ideas on flows to the numbers coming out of his computer, and this opened the subject to scientists and engineers.

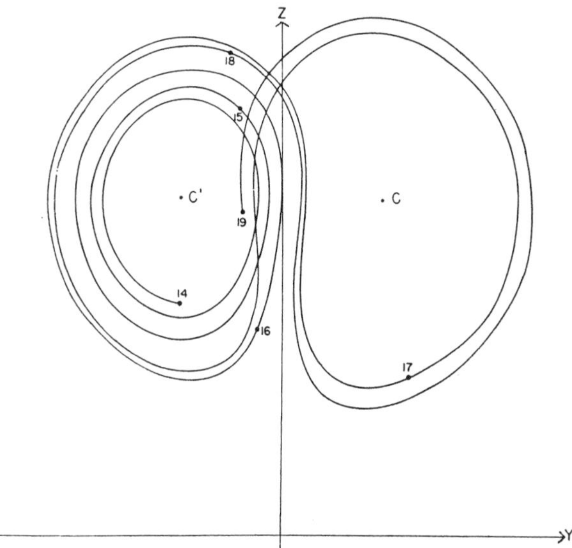

Figure 7.6. Lorenz's breakthrough: this figure shows a hidden pattern not easily discovered in figures 7.4 and 7.5. Here one variable $Y(t)$ is plotted against $Z(t)$, one of the other variables. As the time, t, evolves, each newly computed point is plotted and joined to the previous point by the graph, which then traces out the coiled curves. *Journal of the Atmospheric Sciences* 20 (1963): 130–41. © American Meteorological Society. Reprinted with permission.

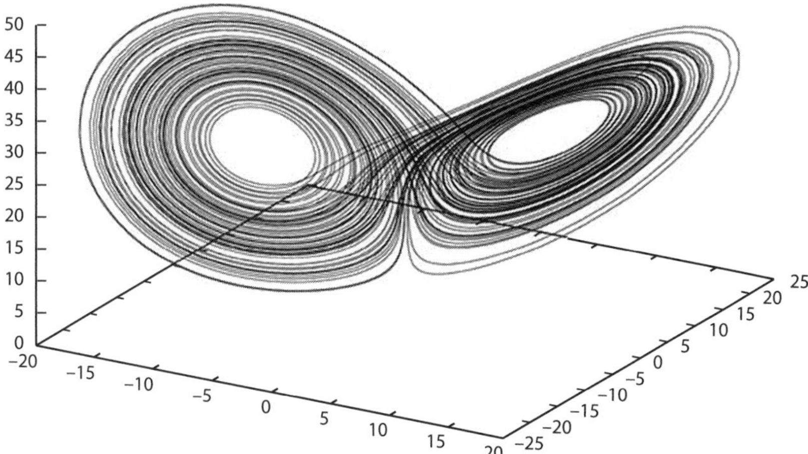

Figure 7.7. The Lorenz "attractor" is shown here by combining the information displayed in figures 7.5 and 7.6 into a single curve, or life history, suspended in X, Y, Z–space. This shows more clearly how all three variables change with respect to each other. The term "attractor" indicates that, although we might expect the curve to wander around randomly (rather like the curve in figure 7.5), instead its path always lies close to two disc- or wing-like surfaces and gradually fills up most of these "butterfly wings."

Their application of Lorenz's ideas on flows led to discoveries of chaos in many widely differing applications.

Behind the Portrait of a Butterfly

Lorenz discovered a pattern, which is the signature of chaotic behavior. So certain systems produce recognizable but not exactly recurring patterns. To understand these new patterns we need some new mathematics.

The principal tool that Lorenz (and Poincaré) found invaluable for studying and understanding chaos was something of a mathematical antique—geometry. Constructing triangles, circles, and straight lines with a ruler and compass was part of the backbone of a good high school education for centuries past. Our following description of geometry is not wasted, as it applies to both the basic static and dynamic states of the atmosphere, and it enables us to see how this ancient tool of measurement and calculation can be used in modern supercomputer simulations of weather and climate, where apparently all that computers do is produce numbers—billions and billions of them.

Geometric thinking has encompassed and underpinned many of mankind's most important developments, yet the geometric principles themselves are often very simple. The ancient Greeks developed geometry as an intellectual activity from about 640 BC onward. Euclid of Alexandria (circa 300 BC) is considered the father of geometry, and his work *Elements* remains a classic text. When geometry started as, literally, the science of measuring the surface of the Earth, there were many practical applications, such as the need to tax farming at rates proportional to the surface area that produced crops. Because the areas involved in these applications were so much smaller than the whole surface area of our planet, the (ideal) Euclidean geometry that developed was thought of as lying on an infinite perfectly flat plain (or "plane," to a mathematician). By 250 BC geometry had been applied to astronomy, where motions of the planets were described by circles; to ratios and patterns that explain harmonies in music; and to surveying and building, including design principles of symmetry and the "golden ratio" as used by many artists.

Looking around us today, we see geometry playing a key role in our lives. The more obvious areas are architecture (see figures CI.10 and

CI.11), interior design, and satellite-based navigation used on land, sea, and in the air; but there are also far from obvious developments, such as the software for medical scanning equipment. Above all this practical benefit, geometry provides us with a way of *reasoning*, as with all mathematical disciplines, and, as such, it is more than just a convenient language for describing the universe.

To appreciate how and why geometry plays a central role in modern mathematical physics, we first have to understand why geometry was, in fact, displaced from its prominent position in mathematics by the new ideas that were introduced during the Renaissance. Until the beginning of the seventeenth century, calculations largely relied on geometric construction—a tangible process of graphical arithmetic—which was essentially restricted to the space we live in. The new era of intellectual development after 1650 ushered in algebra and calculus, which became the new, powerful, and efficient method for solving practical problems in virtually any area of mathematical science. Mathematicians then followed an almost universally accepted agenda of removing geometry from its hitherto eminent position. After nearly two thousand years of proof by Euclid, the pendulum of accepted wisdom swung to the opposite extreme. This program was carried out because the potential of a new and clever way of replacing geometrical figures by abstract symbols became overwhelming.

The leading figure in our story here is the same man who enunciated the principles of good scientific practice that we described in chapter 2, René Descartes. In 1637 Descartes threw away his proverbial ruler and compass. He realized that any problem in geometry could be reduced to one in which only the notions of the lengths of certain lines and the angles between these lines were required. There were rules to find the lengths of the lines in terms of *coordinates*. Anyone who has used the index and grid reference in a road atlas to find a location on a map has used the essential ingredient of Descartes's work. The idea is that a point in space can be specified by a set of numbers denoting its position. For example, any point on the Earth's surface can be indicated by its latitude and longitude. Then distance between such points on a flat map is worked out quickly by using a ruler and the scale, or equally well by formulas (such as those inside modern GPS systems) involving the coordinates.

In his only published mathematical writing, *A Discourse on the Method of Rightly Conducting the Reason and Seeking Truth in the Sciences*,

Figure 7.8. Descartes (1596–1650) was a curious character. While still at school, his health was poor and he was allowed to rest in his bed until 11:00 in the morning—something he continued to do for the rest of his life because he considered it essential to his creativity. Descartes's mathematics has provided us with the modern basis of all calculations and computations. In 1649 Queen Christina of Sweden invited Descartes to Stockholm but insisted that he get up at 5:00 each morning to teach her how to construct tangents to curves. Descartes broke his habit of a lifetime, and, within a few months of watching every sunrise, he died at the age of fifty-four. He was buried in Stockholm, but seventeen years after his death his bones were returned to France and reinterred in Paris. The French nation honored Descartes by renaming his birthplace in Touraine after him. Frans Hals, *Portrait of René Descartes*, ca. 1649–1700, Louvre Museum, Paris.

Descartes wrote about mathematics, optics, meteorology, and geometry. In *Discourse*, he laid out the basic scheme for what we now call analytic, or coordinate, geometry, and in a one-hundred-page appendix called "La géométrie," he set out to achieve the construction of solutions to geometric problems using algebra. This work was the first step toward an entirely new and powerful technique in mathematics and was destined to become one of the most influential publications in the subject.

Descartes introduced and applied algebra to geometry not only to represent a point by a set of numbers but also to represent lines and curves by equations. Reasoning now involved rules for doing arithmetic with symbols rather than numbers. This meant that an infinity of cases of solving an abstract equation could be checked at once, rather than finding a solution in terms of numbers as a particular case. Further, checking someone else's calculation was straightforward compared to checking someone else's geometric construction. This meant that, for the first time in history, problems in geometry were reduced to procedures that involved numerical actions (that is, algebraic manipulation) on lists of numbers.

The importance of this work lies in the fact that it established a correspondence between geometric curves, or pictures in space, and algebraic equations, which could be solved by manipulating symbols in a manner that was subject to repeatable routine or rules. This was a crucial development for programmable computing machines, which hundreds of years later would carry out such solution techniques speedily and accurately. Today we are reversing this regime change. Machines routinely operate with rules on very long lists of numbers—we want to discover or know if there is significant geometry behind these extremely lengthy calculations. Then we can use the geometrical structure to improve the method of calculation by capturing the important qualitative features of the original physical problem to which the geometry relates. The new paradigm is that truth in numbers is revealed by pattern, or meaningful geometry.

In "La géometrie," Descartes introduced another important innovation: the exponential notation for powers, so that "y times y" is written y^2. More subtly, he shows how to multiply and divide these variables. Previously, y was thought of only as a length, so y^2 had to be area. But Descartes now made this process totally abstract, so that arithmetic (and later algebra) became totally independent of any physical basis. Thus, the dimensions of quantities become purely abstract, which meant that general algebraic expressions could be written down involving different powers of variables, such as $y^2 + y^3$. In terms of dimensions, this would not be allowed—how can we meaningfully add an area to a volume, as these products had always been characterized? But Descartes had lifted calculation from these physical interpretations to have meaning in terms of manipulations of abstract symbols. In this sense, a modern computer has no intrinsic relation to the real world but is an abstract calculating machine. A computer has no sense when y is a length, or a cost, or a blood pressure—it just follows its rules laid down in the programs and leaves the interpretation of the result to the user. So when the computer calculates variables that move the air around, for instance, we users need to be reassured that the program respects the laws that represent the physical reality that lies behind the calculations.

Consistent with the notion that the computer gives no physical significance to the weather pixel values is the idea that the key mathematics controlling the weather evolution should be independent of whether we measure distance in miles or millimeters, or speed in knots or meters per second, and so on. Is pressure best measured in inches of mercury, or in millimeters, or even as fractions of the total atmospheric weight at mean sea level? These considerations imply that we can rescale at will, or we can change the way that we measure all our weather pixel variables. This includes changing reference levels from which we measure. Yet we must always get the same final "answer." The actual weather cannot depend on the way that we measure it. These rescaling ideas support the views presented in chapter 6, where Charney identified the dominating terms in our weather equations. The notion of scale in geometry allows us to identify important qualitative features of the physical world—their basic form—independently of how we measure things.

In the last quarter of the seventeenth century Newton and Leibnitz developed the ideas behind calculus, and algebra played a central role in the development of this hugely important technique. By the middle of the eighteenth century this abstract movement was certainly the vogue. The French mathematician Joseph-Louis Lagrange produced a monumental classic, *Méchanique analytique*, in 1788, in which he boasted that the reader would find no diagrams or figures cluttering the pages: geometry was to be eschewed, and mechanics was to be explained by analysis, the new mathematics of Newton, Leibnitz, and their contemporaries. Paradoxically, through this very publication, geometry would creep back into mathematics and now permeates modern physics. But the new geometry had to be abstract and not tied to the space in which we live.

The key idea introduced by Lagrange in his analytical treatment of the theory of motion subject to forces was the notion of a so-called *generalized coordinate*. Whereas coordinates had always been thought of as measurements used to describe actual positions or configurations in physical space, Lagrange used coordinates more abstractly. He also elevated the notion of velocity to the status of a coordinate. For instance, Lagrange would use angles instead of positions when it was more convenient, as we did in chapter 6 when describing the double pendulum. This means that if we want to describe the solar system in the so-called Lagrangian framework, then we have to specify not just the position of all the planets (their configuration) but their motion too. In other words, Lagrange's description amounts to a mathematical "snapshot" of the configuration *and* motion of a system at any moment in time. At any one time, the position and velocity, Lagrange claimed, specify all we need to know about the state of a system or body governed by Newtonian mechanics in order to predict any future position and motion of that system. Generalized coordinates describe an abstract space made up from these variables, which provide a "canvas" for the portraits Lorenz was to discover nearly two hundred years later.

Using Lagrange's state space, we apply geometry not only to the static, unchanging world around us—the world as depicted in a photograph—but also to the changing world, the world as a movie. Just as static forces support the spatial geometry of buildings and bridges and the resting state of our atmosphere, so dynamic or changing force patterns associated with movement in state space are described by a new geometry of motion involving space and time.

It is somewhat ironic that Lagrange, having created the idea of generalized coordinates in pursuit of his program to remove geometry from mechanics and physics, provided us with the perfect arena for demonstrating that more modern ideas of geometry are an integral part of mathematical physics. This remarkable alternative use of Lagrange's thesis began with the work of an Irish prodigy, William Hamilton.

Hamilton pursued research at the interface of mathematics and physics. Two hundred years earlier, Pierre de Fermat (of "Last Theorem" fame) had discovered a principle that states that light always travels by the path that minimizes the time taken for its journey. That is, whether traveling through empty space in a straight line (which is the

Figure 7.9. Sir William Rowan Hamilton (1805–65) was born in Dublin and, before he was a year old, he was placed in the care of an uncle who was a clergyman. Through his uncle, William received a thorough education in thirteen languages and developed a life-long interest in poetry—in his later years becoming a friend of William Wordsworth. Hamilton's interest in mathematics began when he met Zerah Colburn, the American prodigy in mental arithmetic. Enthused by a display of Colburn's abilities, Hamilton obtained a copy of Newton's *Arithmetica universalis* from which he learned geometry and calculus. He then read *Principia* and followed that up with Laplace's *Méchanique céleste*.

shortest distance between two points) or traveling through some optical medium—such as water, which may bend the rays—light travels to its destination as quickly as it can. Hamilton was able to take Fermat's principle and produce a general mathematical description of optics. He formulated his ideas at the age of just seventeen in a paper presented to the royal astronomer of Ireland. Within ten years of publication, this paper was recognized as significant in two ways: first, Hamilton, at the age of twenty-seven, was the royal astronomer for Ireland, and second, he had realized that the same ideas could be applied to the mechanics of motion.

Hamilton's ideas are embodied in what we now call the geometry of *state space*. State space is a very practical idea: put most simply, it is a way of representing *all possible* states of an object or system. Remember that the word "state," in the examples of mechanics that Hamilton (and Lagrange) considered, refers to the position *and* the motion of something (mathematicians call this "phase space"). For example, if we take a picture during a game of football or soccer and send this picture to a friend who is not watching the game, they will be able to see the position or configuration of the players at the moment the picture was taken, and they might just be able to imagine where all the players and the ball were heading at that moment too. But usually there is doubt as to what happens next because the actual motion is not shown. If we could take the same picture in state space, then all the positions as well as the motion of the players at that instant would be recorded—and Lagrange showed that henceforth there would be no doubt. To a mathematician, state space is the abstract but precise way that we record position and motion (strictly as momentum, rather than velocity in mechanics): where we are *and* where we're going. The price we pay is that this state space ends up having too many dimensions to visualize directly. For one object moving in physical space, we need six coordinates. Hence, the geometry we shall introduce becomes somewhat abstruse.

To grasp the essentials, we first focus on a mechanical problem in which the state space is the simplest possible and we can easily visualize it—that is, it is only two-dimensional, with one coordinate for the position and one for the motion. We consider the example of the motion of a simple pendulum, such as that on a grandfather clock. We begin by drawing axes denoting the position and velocity of the pendulum relative to its state of rest. The numbers on the horizontal axis are a measure of the position in terms of the angle that the pendulum makes with the vertical, and the numbers on the vertical axis measure the motion of the pendulum in terms of the changing angle, that is, the rate of rotation of the pendulum. These axes map out the state space, and we sketch an oscillation of the pendulum in this space in figure 7.10.

We now imagine another experiment in which we start the swinging motion by moving the pendulum further to the left and then releasing it from rest. This time the pendulum swings on a bigger arc, and when we plot the motion in state space, the new orbit is a larger circle. Note that if we take a photograph of the pendulum when it reaches the hanging position and show it to a friend, they cannot tell which experiment we have photographed. But the state space picture is precise. It tells us exactly which experiment was photographed. We can repeat this for many different starting positions and obtain a set of life histories. Here each life history is a circle in state space. As time elapses, we imagine a point representing the position and velocity of the pendulum at that moment moving around such a circle.

The complete picture, where we include all such life histories in state space, is the *flow* of the pendulum (a bit like the idea behind the game of Poohsticks, and the path of Hurricane Bill, mentioned in chapter 1, and also the Lorenz attractor shown earlier in this chapter). Here the flow

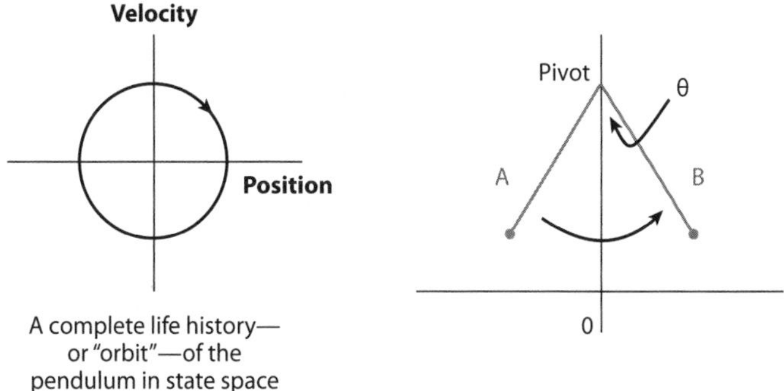

Figure 7.10. The diagrams show how we create an orbit in state space (on the left) that represents the swinging motion of a simple pendulum (on the right). In state space, the horizontal axis represents the angular deviation, θ, of the pendulum from the vertical, while the vertical axis shows the pendulum velocity, $d\theta/dt$, in terms of this angle. We plot larger circles when the pendulum swings further and faster. Periodic behavior is marked out in state space by simple closed curves, which are frequently sought in experimental or numerical data (just as Lorenz did).

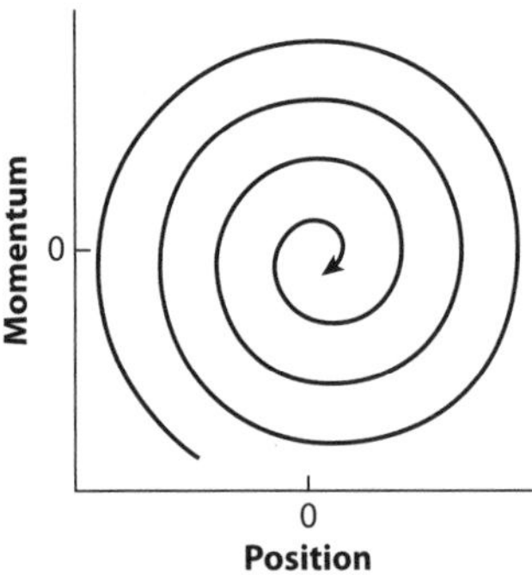

Figure 7.11. The spiral is a life history for a pendulum with friction. Each point corresponds to a state of particular position and velocity, and the succession of points around the curve corresponds to the evolution of the system with time. As friction reduces the maximum angle of swing and speed of the bob, the state space plot spirals into the equilibrium position at the center.

is the family of circles whose centers lie at the origin, and radii vary in proportion to the amplitude of the swing.

Of course, we know that a real pendulum will gradually swing less and less until it stops altogether and hangs motionless once more (as when the grandfather clock is allowed to wind down). This behavior is represented in state space by an orbit that spirals in to the center of our diagram in figure 7.11, where the center represents the state of no deflection and no motion. Similarly, this new flow will be a whole family of spirals that would look rather like water swirling and flowing down a plughole.

The utility of state space comes from our ability to grasp the *qualitative* features of flows from these families of curves without necessarily understanding the detail—a "top-down" view. Families of solutions denote the different possibilities allowed by the laws of motion starting from nearby initial conditions for a given physical system. The amount by which these life histories focus or move apart tells us about the sensitivity of the system.

The Lorenz attractor of figure 7.12 displays the order present in a chaotic system, where the figure shows ensembles of life histories. Each solution is distinguished by starting from different initial conditions. We see that each solution broadly follows the same path, as did those calculations for Hurricane Bill shown in figure 1.19. Each life history never intersects itself, so the system never repeats its state; that is, the system is not periodic or exactly recurring in its pattern.

In figure 7.12a, b, and c we illustrate the spread of computer integrations from different ensembles of initial conditions. The different initial points can be considered as estimates of the "true" state of the system

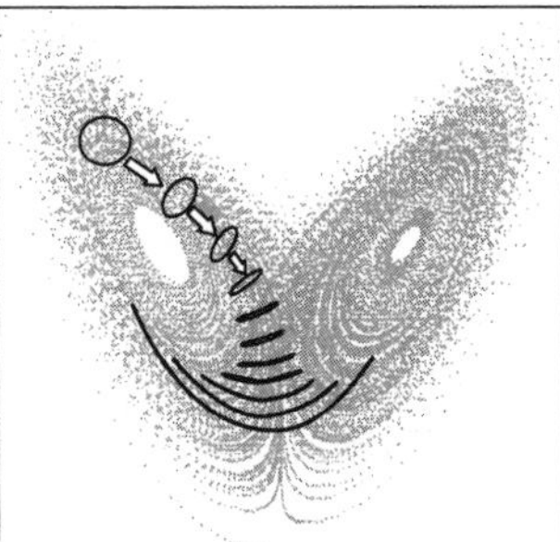
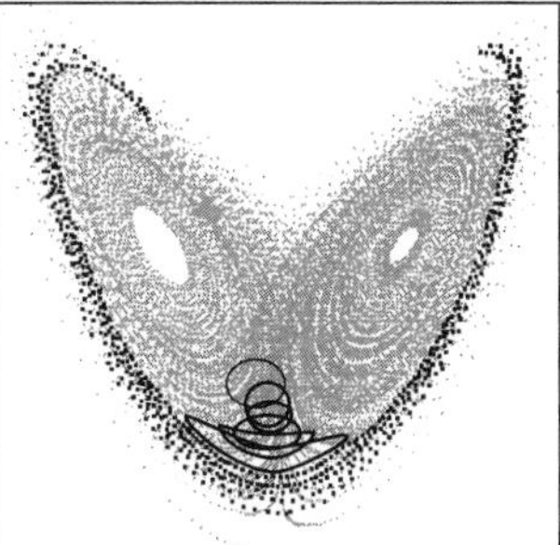

Figure 7.12. We show the output from many computer integrations of the Lorenz equations. The Lorenz attractor is in gray. The initial circle of states leads to less and less predictable final states as we change the initial state from that in (a) to that in (c). © ECMWF. Reprinted with permission.

and the time evolution, or life history, of each of them as possible forecasts. Points close together at initial time separate as time increases at different rates. Thus, depending on the point chosen initially to describe the system life history, different forecasts are obtained.

The two wings of the Lorenz attractor in figure 7.12 can be considered as identifying two different weather regimes. Suppose the main purpose of the forecast is to predict if the system will undergo a weather transition, and if so, when. When the system is in an initial state, such as depicted in figure 7.12a, all the points stay reasonably close together till the final time. Whatever the life history chosen to represent the evolution of the system, the forecast is characterized by a uniformly small error, and a correct indication of a weather transition is given. The "ensemble" of points can be used to generate probabilistic forecasts of weather transitions. In this case, since all points end in the other wing of the attractor, there is a 100 percent probability of transition for these initial conditions in figure 7.12a.

By contrast, when the system starts from a state as indicated in figure 7.12b and 7.12c, the life histories stay close together only for a short time and then start separating and spreading out. While it is still possible to predict with a good degree of accuracy the future forecast state of the system for a short time period, it is difficult to predict whether the system will go through a weather transition in the longer forecast range. In figure 7.12b the majority of the life histories do not have a weather transition. Figure 7.12c shows the results of a more sensitive choice for the initial ensemble than figure 7.12b, with life histories spreading rapidly and ending in very distant parts of the Lorenz attractor. In probabilistic terms, we can predict that there is a 50 percent chance of the system undergoing a weather regime transition, but we cannot say if our particular initial weather state will lead to such a change. Moreover, the spread of final states in 7.12c indicates that there is great uncertainty in predicting the final state of the system. This example shows why much of modern forecasting involves predicting the likelihood or probability of future events rather than trying to find the exact future event. We return to this in chapter 8.

So Euclid's pictorial thinking about triangles, circles, lines, and perpendiculars evolved into abstract operations on lists of numbers. Static geometry in the physical space that we live in became algebra. Then motion was given its own coordinates, and motion became differential equations and algebra. Simple stable motion remains near static or periodic states, while unstable motion winds all over (parts of) state space. But certain "chaotic" motions still remain anchored in state space, near the "butterfly wings" in the case of the Lorenz equations.

Although Lorenz's model has been used extensively as a paradigm for weather, his toy model is actually too simplistic in one rather important aspect. The model does not respect any of the overarching principles we discussed in chapters 5 and 6, principles that underpin the patterns we observe in weather and climate. Charney's remarkable physical insight showed how the equations of weather obey scaling rules, and showed that they also obey conservation principles. The next step is to explain how geometry codifies some of what is otherwise invisible in the storm—the mathematical description of the overarching laws and principles that constrain the possibilities.

Constraining the Possibilities in State Space

Our modern view of geometry captures the truths that appear no matter how we measure or scale the physical variables, or even how we may specifically transform them. Figure CI.10 shows the design of an English church and depicts a feature that is independent of its precise size. One way to think of that feature is to consider the balance of static forces that hold the weight of the structure up. The columns allow us to visualize the forces, and the arches show this force being spread through the building. Scale models of structures, such as buildings and bridges, are tested to see whether they will resist gales and earthquakes. More concretely, the stability of the building should not depend on its size. Similarly, the features, or truths, behind the scientific laws should not depend on the method of measurement of each of the variables. We call this *invariance under rescaling*.

A similar static balance in the atmosphere is shown by cumulus cloud layers on fairly calm days. On a fine spring day, these layers of cumulus clouds can appear much the same at latitudes 25 degrees north—for example, over Florida—as at 60 degrees north—for example, over the Canadian prairies. The point of difference is elevation of the clouds, and the explanation is that the warmer air above Florida has expanded more, and lifted the cloud layer. Visually the vertical scale of the atmosphere changes, but little else changes, as the drawing in figure 7.1 also suggests. Can we find a way to exploit these universal scalings?

Rising moist air creates clouds when the water vapor reaches a critical (cooler) temperature of a few degrees centigrade. This is similar to exhaling our breath on a frosty morning, making clouds of steam. On a fine spring day, the condensation level in the atmosphere is typically between one and three kilometers above sea level, and the level is characterized by a pressure surface (see figure 7.13). Sometimes the condensation level reaches the ground, especially when warm moist air flows over a much colder sea surface. Figure CI.11 shows this frequent occurrence in the San Francisco Bay area.

Bjerknes, in his first Carnegie volume on statics in 1908, addressed these issues with regard to the atmosphere. Bjerknes worked out the benefits of using pressure as a means of measuring height above sea level. Because pressure depends on the weight of the atmosphere above, it does not matter how expanded (by warming) the gas that makes up the air is. It still has the same weight. Pressure surfaces follow the height scalings shown in figure 7.1 and allow us to focus on the (mainly horizontal) movement of air—this is what the use of height was doing in chapter 6.

Moving on from statics, we resume our quest for a geometric description of the weather as a movie—a geometric structure within state space—by returning to one of the simplest nonlinear chaotic physical systems, the double pendulum introduced in chapter 4. We have just seen how the periodic motion of a simple pendulum is described by a

Figure 7.13. Cumulus clouds on a fine day appear much the same over much of the Earth's surface. What changes systematically with latitude is their elevation, an example of rescaling with respect to height. © Robert Hine.

circle in state space, as shown in figure 7.10, and we now show how the chaotic life history of a double pendulum always lies on a particular surface in its state space. The point we make is that constraints are crucial if we are not to muddle different families of chaotic life histories.

The state space of the double pendulum with smoothly joined rods shown in figure 7.14 (above left) is four dimensional; that is, the angles A and B in the figure, and the rates of change of A and B, give the swinging motion of the rods, illustrated by the dashed arcs. Together these give the position and motion of the rods; that is, the "state" of the rods. So it is impossible to visualize this state space in any conventional way on paper or in our (three-dimensional) visual space. These angles are Lagrange's generalized coordinates. The effectiveness of these coordinates is shown in figure 7.14 (below), where we see that the trajectory of the end of the second rod is a curve lying on the surface of a torus, or "doughnut," in the space of the configuratons of the rods (so this is not the complete state space, which also requires the velocities).

It is the mechanical construction of the double pendulum that leads to the existence of the torus in configuration (or position) space: a pivot joins two rigid rods and the end of one rod is pivoted at a fixed location,

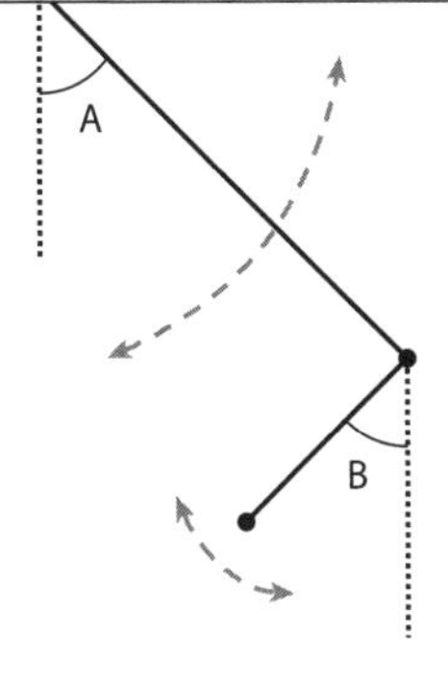

Figure 7.14. Above left, the motion of an idealized double pendulum is shown. The positions of its arms are specified by the angles *A* and *B*. But what is the motion? Which way are the arms swinging? State space would tell us this, but it requires four dimensions. However, we can represent the configuration of the arms at any time as a single point on an abstract surface in a three-dimensional space. Above right, the random life history, shown here by attaching a light bulb to the end of the second pendulum and taking a long-exposure photograph, actually lies on a torus (or "doughnut") in the configuration space of the system, shown below. Because of the physical constraints created by the pivots, no matter how complicated the curve gets, it always remains on the surface of the torus. Above right figure © Michael Devereux. Figure below © Ross Bannister.

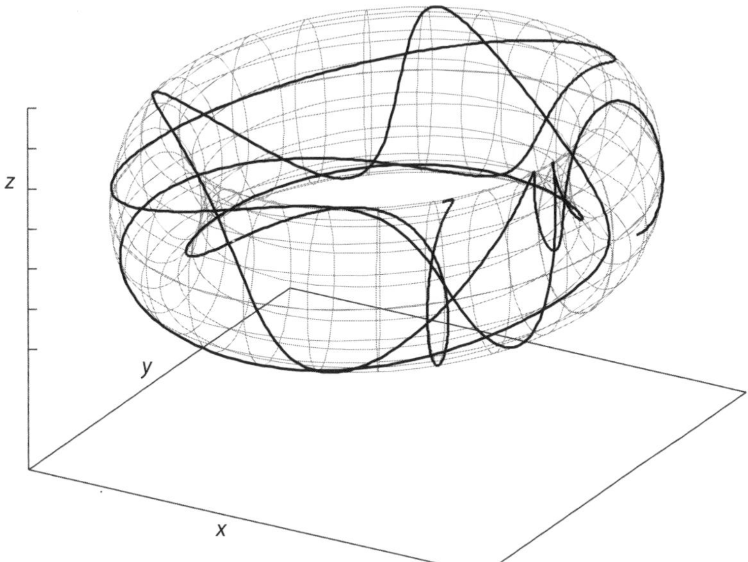

so the only motion is the angular rotation of these rods about their pivots (see figure 7.14 [below]). In other words, physical constraints—the two pivots—confine the motion of these rods when whirling around in space, and it is precisely these constraints that lead to the dynamical trajectories lying on a torus.

However, we still have not got to what ultimately controls the motion of the double pendulum, which is conservation of the total energy of this idealized system. When one of the pendulums is lifted against gravity, it gains potential energy. This potential can be realized when the pendulum is released and it starts moving, gaining motion energy. The sum of the potential and motion energies is always constant for such idealized mechanical systems, since the idealization assumes there is no friction. The conservation of such total mechanical energy provides an abstract constraint in state space. This constraint forces the life histories to lie on nice surfaces in state space rather than entangle as messy collections of orbits would.

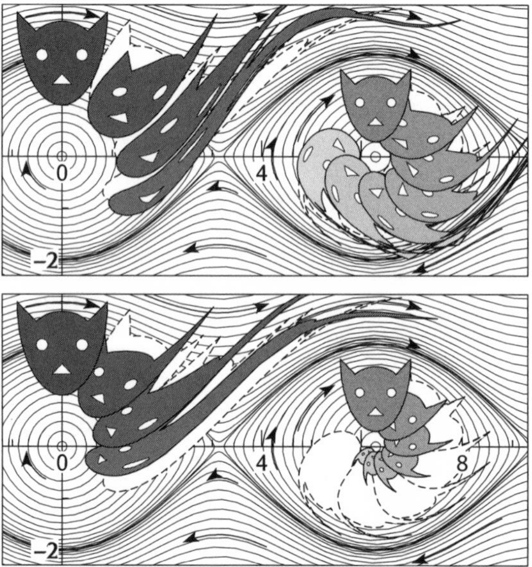

Figure 7.15. An abstract area in the state space of a pendulum is illustrated by the animal's face. The dashed lines (more clearly visible in the lower panel) show exact area preservation. The simulation shown in the upper panel (symplectic Euler) preserves area to a very good approximation, even as the flow distorts, while the simulation shown below (implicit Euler) does not preserve area: the face shrinks relative to the dashed lines within the right hand lobe. Reprinted with kind permission from Springer Science+Business Media: Ernst Hairer, Gerhard Wanner, Christian Lubich, *Geometric Numerical Integration, Symplectic Integration of Hamiltonian Systems*, 2006, p. 188, figure 3.1.

The point of all this is that there emerges an intimate connection between basic objects common to geometry—points, lines or curves, and surfaces—and dynamical behavior when viewed in state space. Points in state space depict the full state of motion of a system at any moment of time; lines or curves depict the life histories, the solutions of the equations of motion (as in the case of the Lorenz butterfly); and these solutions often lie on surfaces, which indicates the presence of some underlying relationship between the variables (relationships that may not be obvious from the physical apparatus, or even from the governing equations). In the atmosphere, hydrostatic and geostrophic balance act as important constraints on the development of weather systems. Further, there are mass and energy conservation laws for air parcels, and we introduced the important conservation law for potential vorticity (PV) in chapter 5. Our aim is to find geometries that describe both constraints and conservation laws in state space. Our belief is that computer algorithms give more accurate descriptions of weather when the algorithms respect these geometries, as the animal's face in figure 7.15 suggests.

Geometry and State Space

The geometry that encodes PV conservation is relatively modern—barely one hundred years old compared to Euclidean geometry, which is more than two thousand years old—so our pursuit of geometry in state space requires more contemporary ideas. Ironically, the one obstacle to the application of geometry to physics that so influenced the likes of Lagrange and his contemporaries in the 1700s was the monopoly Euclid had on the subject. Euclid synthesized the geometric knowledge of idealized figures in the physical world, and his method was to derive all known geometric truths from a small collection of self-evident assertions or axioms. Philosophically, it was argued that Greek geometry was discovering truths about real objects, even if these were taken as perfect in shape. Euclid's program is now seen as a way of stating a small system of basic assumptions and then deriving their consequences. But it took more than two thousand years for mathematicians to realize that geometry is an abstract method of reasoning rather than a way to describe physical reality. And there might be more than one kind of geometry: the geometry of the universe might turn out to be non-Euclidean.

By the 1850s, faced with the realization that a variety of different geometries might be constructed or invented, mathematicians began to search for alternative underlying principles that would quantify the essence of "true geometry." In 1872 Felix Klein of the University of Erlangen suggested a unified point of view. According to Klein, geometry is not really about points, lines, and planes but is about *transformations*. The size of an angle remains unchanged if the angle is translated in the plane, or if the plane is rotated. We may even shrink, or scale, the picture—angles (and ratios of lengths) will still be the same. In fact, all the basic mathematical objects described by Euclidean geometry remain unchanged if translated or rotated: circles remain circles and parallel lines remain parallel, and so on. It is the invariance of objects of interest under certain classes of transformations, Klein argued, that truly characterizes geometry.

Euclid begins with an idealized world of infinitely small points, infinitely thin lines, and perfectly flat planes. Most of his basic assumptions about this idealized system are simple and plausible, but one postulate stands out as being difficult to state and far from self-evident. This is the assumption that two straight lines in a plane will never meet if they are parallel, such as railway lines. How can we be sure that this is true? Real railway lines lie on a curved, not a flat, surface—the surface of the Earth—so we cannot appeal to our experience here. Euclid's successors tried in vain to deduce the parallel postulate from the other assumptions. Lagrange showed that the sum of the angles of a triangle being 180 degrees is an equivalent assumption, but it was becoming clear by the beginning of the nineteenth century that any such attempt was doomed. In fact, there were other geometries where this was no longer true. Examples of such geometries are affine, projective, conformal, and Riemannian; the latter became a cornerstone of Einstein's General Theory of Relativity. Affine geometries allow us to use vectors in problems of mechanics involving forces and their points of application. Projective geometries allow us to map from the surface of a ball onto a plane so we can make flat maps (for example, of the world—more on this later). Conformal geometries introduce the world of complex numbers (and Hamilton generalized these to quaternions later in his life) and complex functions; the mysterious purely imaginary number i (= $\sqrt{-1}$) has a geometric interpretation of rotation counterclockwise by 90 degrees.

Euclid realized that his main technique for proving results about points, lines, and figures such as triangles involved matching shapes. The reader had to imagine lifting, rotating, and moving sideways a rigid copy of one shape to match it to another, usually to establish congruency of triangular shapes, just as in the proof of Pythagoras's theorem (see figure 7.16). Often one shape had to be proportionally scaled so that it exactly fits the other, such as the family of right-angled triangles in figure 7.16, which scale like the circulation cells of figure 7.1.

The Euclidean idea of transformation can be generalized considerably. In Euclid's proofs, rescaling (expanding everything uniformly) was typically the only "stretching" or nonrigid movement allowed. Today we often describe which transformations are allowed by stating what objects do not change during the transformation. If we adopt this view, the invariant quantity in the transformations of figure 7.16 is angle.

A new and quite different geometry is created when, instead of angle, area is taken as the invariant quantity. This is called *symplectic geometry*, after the Greek word for "tangled" or "plaited." It was not until the end of the nineteenth century, when Poincaré was working on the three-body

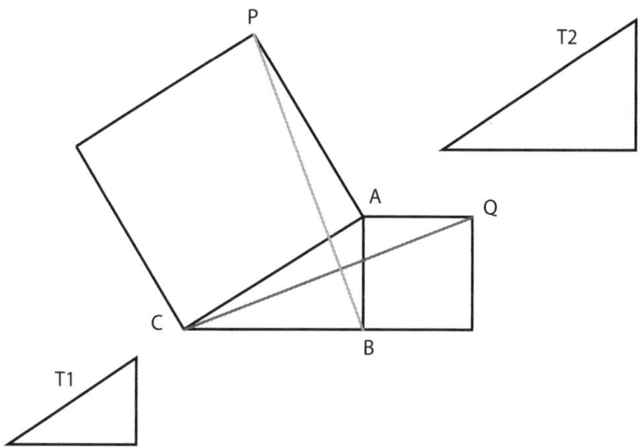

Figure 7.16. The key idea in the proof of Pythagoras's theorem is to show that the triangle with vertices (or corners) *PAB* is the "same" (after lifting and rigidly rotating about the point *A*) as the triangle *CAQ*; hence, these triangles will have the same areas. This identifies the area of the square on the side *AB* with part of the area of the square on the side *AC*. The area of the square on the side *BC* is treated similarly. We also show the right-angled triangle *ABC* scaled down (left, T1) and scaled up (right, T2). This is similar to the scaling of the cells of figure 7.1. Scaling is a vital element of both the analysis of weather laws and the formulation of effective computer algorithms.

problem, that the first steps toward linking Hamilton's treatment of mechanics to symplectic geometry were taken. Poincaré used state space, with its families of life histories, as an essential tool in his assault on the three-body problem, and he became interested in the possible geometrical relationships between the evolution of the position and velocity coordinates. If we examine the position and velocity vectors of a body at a particular location, then they form a small parallelogram in state space, as shown in figure 7.17. Area is calculated from a suitable product of these two vectors. For Hamiltonian dynamical systems, this area must remain the same as the system evolves. This area invariance means that if the position of an object changes, then its velocity must change in such a compensating way as to keep the area constant in state space.

When state space has many dimensions, trying to think about area preservation following a life history path of motion in state space becomes very challenging. So we first consider angle and area preserving geometric transformations in a simple example where no motion is present—that is, in configuration space.

Consider the various ways we can represent the surface of planet Earth as a "flat map." If we want to refer to places in the Indian subcontinent at the same time as places in Brazil (which are on opposite sides of the globe), then a map in a book is usually more convenient. The standard historical transformation from a map on a globe to a map in a book is called a *Mercator* projection, after the sixteenth-century Dutch mathematician and teacher Gerardus Mercator.

As we note from figure 7.18, even though Mercator's projection respects the angles between lines of latitude and longitude, it distorts the size of land masses, especially in the Arctic and Antarctic regions. Can

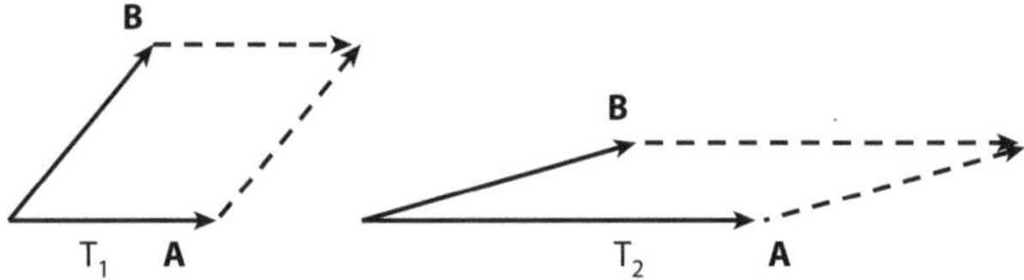

Figure 7.17. The squashed rectangle is the area element defined by the two vectors *A* and *B*. As *A* and *B* change while following a life history in state space, so does the area element, say, from that at time T_1 to that at a later time, T_2. But the amount of area remains the same. Conservation laws involving angular momentum impose this geometric principle in state space. So does PV.

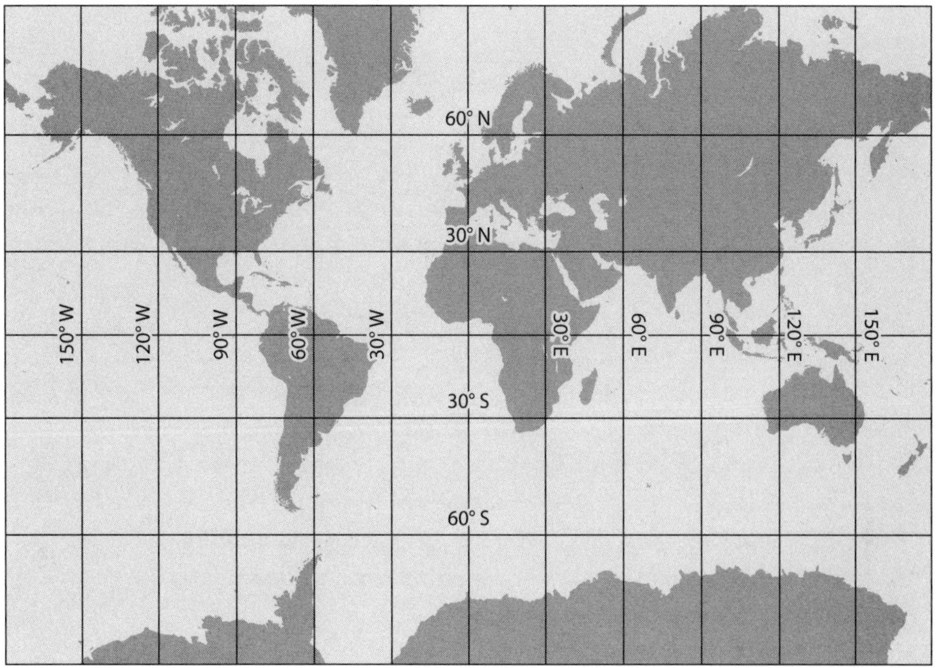

Figure 7.18. Mercator's "projection" shows circles of latitude and longitude on the globe as horizontal and vertical straight lines on the page. Note how this map of the continents enlarges the areas near the poles—Antarctica is enlarged so much that it appears to dwarf North America and Russia. Mercator's projection distorts this area "truth" badly near the poles.

we find a transformation that preserves the areas of continents and countries? The Mollweide projection (named after Karl Mollweide, the eighteenth-century German mathematician and astronomer) sacrifices fidelity to angle and shape in favor of accurate depiction of area. The flat map shows the true area of land masses as measured on the globe, choosing instead to distort their shape, as shown in figure 7.19. Modern maps tend to further modify these projections to respect local country shapes and relations to neighbors—the map projection that we decide is best depends on the use we have in mind.

Having introduced the Mercator (angle-preserving) and Mollweide (area-preserving) projections in static space, we next discuss the related transformations in state space that involve motion. When Charney applied rescaling to the position and motion variables in the basic weather

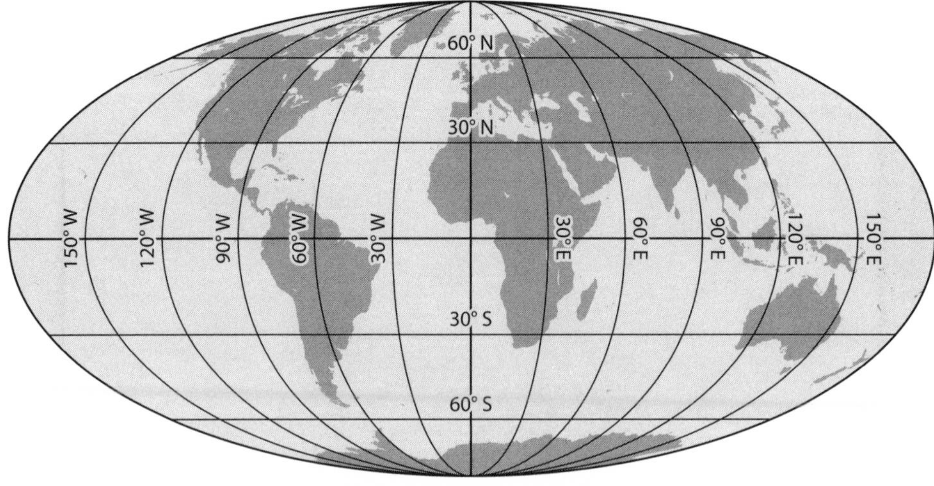

Figure 7.19. The Mollweide equal-area projection gives countries their correct "size" in terms of area but significantly alters their shape when they are near the boundary.

equations of chapter 2, he identified geostrophic motion of the hydrostatically balanced local states as the next most important effect after the hydrostatic pressure force balance that occurs when there is no motion. Thus followed, as we saw in chapter 6, the successful computation of simplified, idealized weather often described as QG evolution. And this QG weather follows the time-honored tradition advocated by Bjerknes: it respects a circulation theorem and has a conservation law for PV. Preserving PV turns out to be related to symplectic, or area-preserving, transformations.

A transformation that preserves area is actually more flexible than a transformation that preserves angle. To see this, watch a drop of cream put carefully onto the surface of a cup of coffee, and then stir the coffee gently round the cup, as shown in figure 7.20. The patch will become distorted, but if the stirring is done appropriately, the patch will preserve its area even when thin filaments develop, as shown in figure 7.15. This transformation of the fluid surface is approximately symplectic. There is a famous example in modern geometry that makes this point very apparent—it is called the *symplectic camel.* Using the symplectic transformation, any camel in symplectic geometry may be elongated and, hence, thinned sufficiently to actually pass through the eye of a needle! We contrast this with the more rigid transformations of a camel allowed

Figure 7.20. Gently stirring the surface of a cup of coffee with a floating patch of cream produces a thin spiral filament of cream. Even though the cream is stretched, it still has the same volume and, hence, the same projected surface area when the thickness remains the same. © Eric Chiang / 123RF.COM.

in angle preserving geometries. Thus symplectic transformations do not respect rigidity.

The importance of area-preserving geometries is better appreciated in meteorology for two reasons. Observed flow patterns in the atmosphere are often complicated. As an example, suppose two air masses with different properties (for example, temperature and humidity) flow around each other and perhaps mix to some extent. We need to be able to identify the "thumbprint" of such processes in the computer models—simply hoping that the billions of calculations add up to something realistic is often insufficient and frequently unreliable. The simulations need to be constrained to represent both the larger patterns and the conservation of heat and moisture on air parcels being transported in the flows.

The Rossby memorial volume contains figure 7.21, which shows how air in an initially uniform checkerboard grid flows and distorts over a period of time—for example, during a developing cyclone. Since the air mass transports moisture, accurate rainfall prediction requires accurate parcel mass transport. Here we think of moisture being transported by the wind just like cream being transported by the moving coffee surface. The preservation of parcel volume is achieved by conserving parcel area when the appropriate thickness coordinate is kept constant. Preserving air mass is more important in getting the amount of transported water correct than knowing its precise shape. If the computer code can be designed to trace the flow of air using symplectic transformations on constant mass thickness surfaces, then the accuracy of the total simulation will be improved, sometimes greatly. In the next section we explore the conservation of PV on moving air parcels.

Figure 7.21 is a drawing from the 1950s. Today we have models capable of simulating the filamentation of flows with considerable accuracy. One example, called contour advection, is illustrated in figure CI.9.

The pieces of our jigsaw are ready to be assembled. We combine a transformation from wind, temperature, and pressure to PV with a new transformation from height to potential temperature as the vertical coordinate—this is to get the mass to symplectic area relation more accurate. Then we arrive at a mathematical description of atmospheric flows that is most naturally expressed in terms of area-preserving geometry.

The Global Picture

The Mercator and Mollweide map projection examples show two geometries that we focus on, the scaling and area-preserving geometries. Scaling geometry has already been in action in chapter 6 (figure 6.9), where we discussed Charney's first identification of the largest forces for typical weather motions. When these largest forces dominate the acceleration terms in the momentum equation, we called this hydrostatic and geostrophic balance. These scalings help us explain the decrease in the scaling height of the atmosphere as we move from the equator to the poles—the decrease occurs whether there is atmospheric motion or

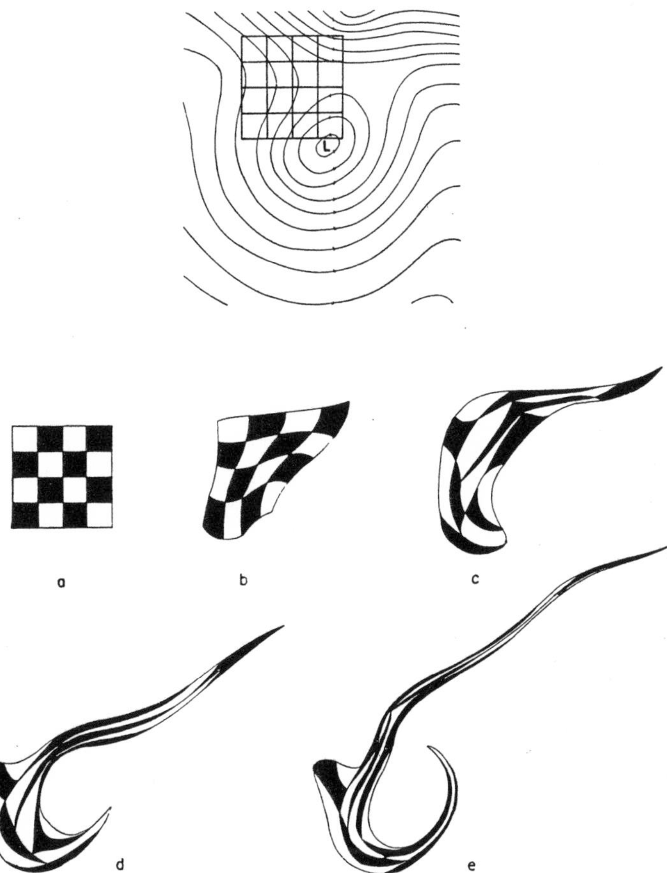

Figure 7.21. The wind associated with a typical low pressure system L (or cyclone) transports a black and white 4 × 4 checkerboard placed in a critical region of the flow. The grid starts at the time (a), and proceeds through times (b) to (e). When this happens to air masses that are carrying moisture, how do we ensure that the computation preserves the moisture in the filaments, given that we only remember values on fixed pixels? This greatly affects our ability to predict rainfall. © 1959, Rockefeller University Press. *The Atmosphere and the Sea in Motion*, edited by Bert Bolin, p. 34.

not—and helps us understand the average wind movement around the pressure contours in the middle and upper troposphere.

These qualitative aspects of the structure of the atmosphere are incorporated into computer simulations by making use of a relatively simple transformation of variables. Rather than looking at weather pixels fixed at the same height above sea level from the equator to the pole, could we find transformations of our pixel quantities to a temperature-influenced height so that each weather pixel represents about the same amount of atmosphere? Then blankets of cumulus cloud layers in these new "height" variables would have approximately the same "height" value at all latitudes. The computer calculations in these variables would then focus on the essence of the changing atmospheric state rather than first spending a lot of time recalculating the basic local static balance state. Such transformations are known in meteorological parlance as changing to isobaric (equal pressure) coordinates, and we focus on the potential temperature as a weather pixel variable, not ordinary temperature. In isobaric coordinates, the atmospheric structure sketched in figure 7.1 would have nearly the same height anywhere from the equator to the poles.

From the viewpoint of an air parcel, the potential temperature (that we introduced in tech box 5.2) summarizes the total thermal energy that the air parcel possesses, and thermal energy is important because it can be transformed into motion energy to speed up the winds. In tech box 7.1, the PV formula of chapter 5, on page 180 is extended to one where temperature changes in the atmospheric motion are now important.

Tech Box 7.1. The Backbone of Weather II: Potential Vorticity in Isentropic Coordinates

We extend tech box 3.2, where temperature effects were not important, to an atmosphere where the potential temperature now matters. The expression for PV for the flow of an air parcel in which density and temperature vary is given by the equation

$$\mathrm{PV} = \frac{1}{\rho}\zeta \cdot \nabla\theta.$$

Here ρ is the fluid density, ζ is the total vorticity (the sum of relative and planetary vorticity—see chapter 5), and $\nabla\theta$ is the gradient of the potential temperature θ. For many atmospheric parcel flows, the parcel PV is constant during the motion. Calculating $D(PV)/Dt$ directly from the equations of chapter 2 is a testing exercise for a student of meteorology.

How does this definition of PV relate to the one we gave in chapter 5? In this tech box we describe how to transform between various expressions for PV—a procedure that dynamical meteorologists use to gain insight into the detailed pixel interactions in state space.

The transformation to isentropic coordinates is accomplished by replacing the vertical coordinate, z (height, shown on the left-hand vertical axis in figure 7.22), with the surfaces of constant θ. Then PV is written

$$PV = -g(f + \varsigma_\theta)\ (\partial p/\partial\theta)^{-1},$$

where g is the acceleration due to gravity, f is the Coriolis term, ς_θ is the relative vorticity on a constant θ surface, and p is pressure. (The partial derivative of p is expressed consistently with θ being an independent variable.) Now clear analogies with the constant temperature form given in chapter 5, on page 180, can be made, as described in the text below.

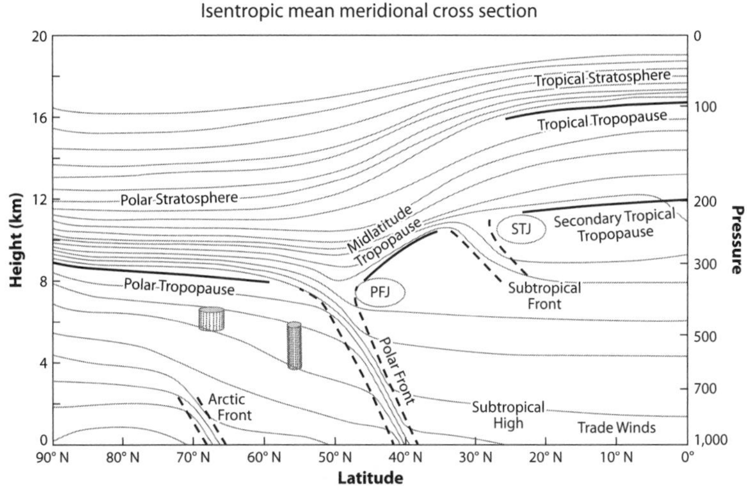

Figure 7.22. A typical cross section of surfaces of equal potential temperature is shown in a vertical section of the atmosphere extending from the North Pole to the equator. The various (annually and zonally averaged) global fronts, jet streams, and tropopauses are shown in the troposphere, between the Earth's surface and the stratosphere. A location for a column of air bounded by two θ surfaces is shown at about 67° N.

Before we discuss PV and θ transformations, we note that, through a combination of the first law of thermodynamics (conservation of fluid parcel energy, which is expressed most simply by using θ) and the horizontal momentum laws, PV is only changed by diabatic heating (such as latent heat released from condensation of water vapor) and "frictional processes" (such as ensue from the movement of the winds near hills and convective plumes). Put otherwise, if these processes do not occur so that the flow is frictionless and adiabatic (which means that θ is constant on the air parcel as it moves), then PV is *conserved* following the motion: in symbols, $D(\mathrm{PV})/Dt = 0$. This conservation law holds to a good approximation in the upper and middle regions of the atmosphere, which are important for the formation of cyclones.

With this conservation principle in mind, we return to the analogy described in chapter 5, where the temperature and density were assumed constant: a spinning ice skater with his arms spread out laterally can accelerate his rate of spin by contracting his arms. Similarly, when a broad, spinning column of air in cross section contracts, the rotational air speed must increase to maintain PV. Conservation of air mass results in a vertical stretching of the column as well.

How do we spot the conceptual simplicity of the convergence/spin increase relation just described in terms of the tech box definition of PV? The secret is to transform the PV so that potential temperature, instead of height, is used as the independent vertical coordinate. These so-called isentropic (or equal θ) surfaces are shown in figure 7.22; two possible locations for a rotating mass of air trapped between the same two θ surfaces are also shown hatched.

Consider such a disc-like column of air in an adiabatic flow. The top and bottom of the column will then be constrained to lie on these θ surfaces as the column moves. The dynamics of the column are governed by the conservation law for PV, and the only quantities that change are pressure, relative vorticity, and the Coriolis term. The stretching of the vertical extent of the column in physical space is given by the change in pressure (which is given on the right-hand vertical axis of figure 7.22). Such changes in vertical extent must be compensated by changes in total vorticity ($f + \varsigma_\theta$). This is the same principle described in chapter 5, which governs the initiation of a Rossby wave in a westerly air current moving across the Andes. Plotting PV using these isentropic coordinates is illustrated in figure CI.12, where we show computer-generated views of the atmosphere from well above the North Pole. Figure 7.23 gives a vivid image of the northern polar vortex splitting in the lower stratosphere for the period January 2009.

Transforming PV to isentropic coordinates enables us to think about area-conserving geometries on θ surfaces. The salient reason for using symplectic geometry is that it turns out to be the natural mathematical language for studying Bjerknes's circulation theorem. Bjerknes did not know of this geometrical interpretation in his studies of atmospheric and oceanic flows because the subject of symplectic geometry was in its infancy at that time; it was one of the topics that received a considerable boost in mathematical analysis as a result of Poincaré's work.

Today we have in place the mathematics necessary to express Bjerknes's conservation principle in abstract geometrical terms in weather state space. The virtue of pursuing an abstract approach is that it opens the door to an effective way of constructing better pixel rules. The physical principles embodied by the circulation theorem are then embedded in the weather and climate models, thereby improving our ability to compute a forecast. This is an ongoing topic of present-day research. Again, accurate air mass transport in the computer model leads to more accurate prediction of volcanic dust or ozone (as shown in figure 7.23), as well as moisture movement, cloud development, and rainfall processes. Pursuit of this goal takes us well beyond Bjerknes's original vision.

Bjerknes used the conservation of the flow of an ideal fluid around a closed curve to explain many apparently differing fluid flow phenomena in nearly horizontal shallow layers of the atmosphere and oceans. This one law ties down the motion of air and water in such a way that these fluids are not at liberty to flow in a truly random fashion. This conservation law for such suitably defined fluid-flow circulation lies at the very heart of a modern geometric view of our atmosphere's motion.

In figure 7.24, we depict an air mass and a contour around which circulation, indicated by the larger arrow, is calculated. We show the outer path being broken into smaller loops on pixels; this construction is quite general and does not depend on any special properties of the outer loop. The point is that as we go to smaller and smaller pixels and loops, we arrive at the areas that are part of symplectic structure. Knowledge of symplectic geometry is needed to build the circulation theorem into the computer codes. Most importantly, we do not wish for random, typically inconsistent, errors to develop in the billions of arithmetical

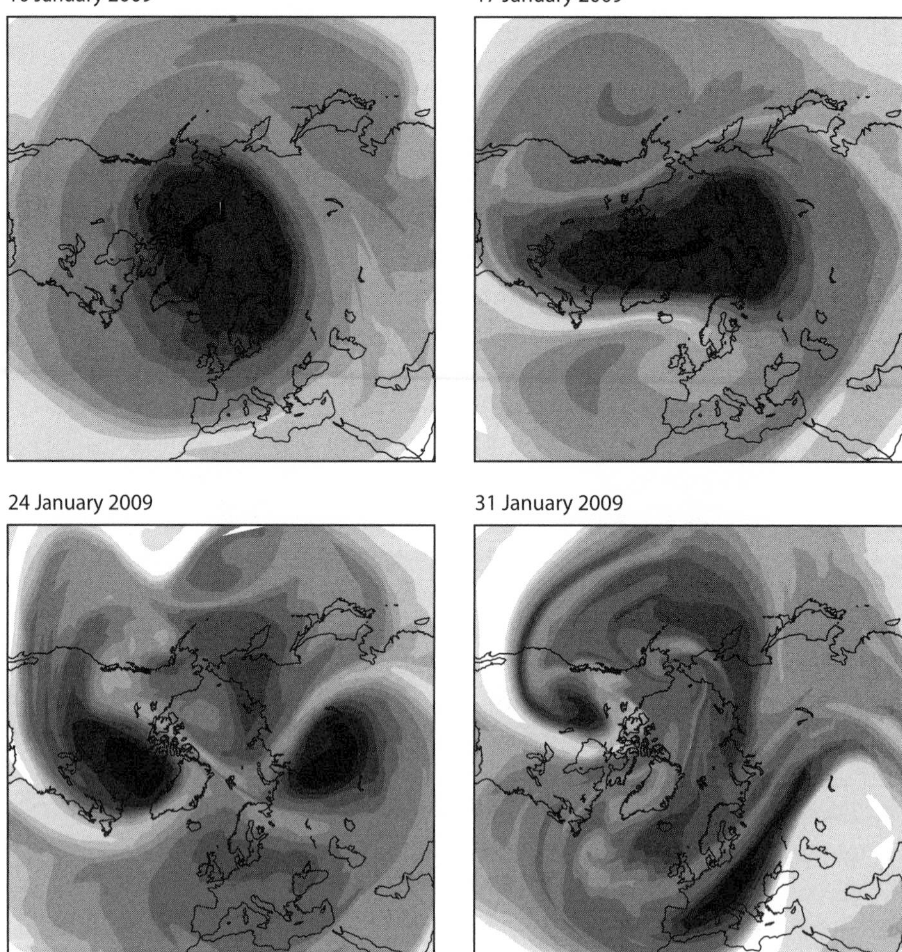

Figure 7.23. Contours of ozone are plotted in the lower stratosphere above the Arctic ocean. In January 2009 a swirling polar vortex splits and wraps in a complicated manner. (ECMWF analysis of stratospheric ozone at 10 mb) © ECMWF. Reprinted with permission.

calculations that happen during the supercomputer simulations. The unavoidable errors in computer modeling the original equations need to be controlled by the conservation principles.

Whether we are interested in the motion of planets in the solar system or the motion of a cyclone or a hurricane in the Earth's atmosphere,

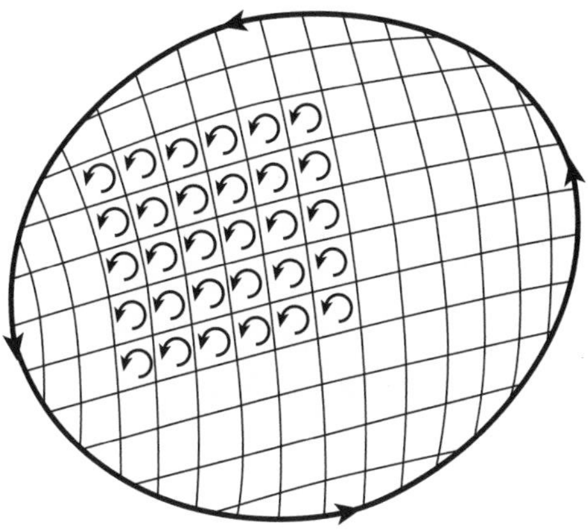

Figure 7.24. The individual swirls on computer pixels add up to a larger swirl on a fluid parcel, ensuring that the circulation theorem holds in the weather that the pixels are trying to represent. The computer rules on each pixel need to be defined so that the internal cancellations on neighboring pixels exactly add to the correct circulation around the larger air parcel. Imposing a geometric view on the billions of individual arithmetic and algebraic calculations helps to do this more accurately.

the qualitative features of such motions are usually consequences of configuration constraints and conservation laws. Scaling allows us to identify dominant force balances, and to make appropriate perturbation approximations, as described in chapter 6. These enable us to cut the Gordian knot of nonlinear feedback. Conservation laws then control the flow possibilities and keep the calculations nearer to "real" weather. The mass, moisture, and thermal energy of air parcels are strictly controlled as they blow about in the wind. Further, the swirling wind itself cannot flow randomly but must respect circulation or PV laws. Modern weather agencies use algorithm and pixel choices that satisfy these rules accurately enough to usually provide successful prediction of atmospheric flow for several days. The next day's weather is usually very well predicted at any place on Earth. In figure 7.25 we show a computer-generated satellite image of ribbons of cloud ending in swirls about low-pressure weather systems. Superimposed are contours of PV, which show the coherence of the computer forecast.

Symplectic geometry is the mathematics that describes the circulation theorem in state space. The "plaited together" of the word "symplectic" makes it an apposite description of the "delicate thread," alluded to by Bjerknes, that ties the air motion to the physical processes involving heat and moisture. Symplectic geometry shows us how we might yet encode this circulation theorem in precise mathematical terms within the detailed calculations at the heart of modern weather forecasts. So

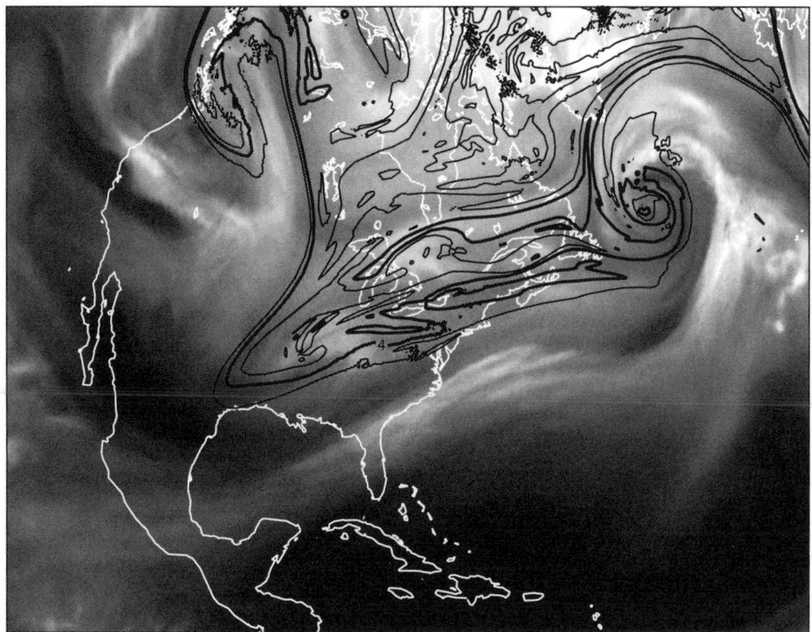

Figure 7.25. Bringing everything together: the contours show PV computed on the 315K surface, and the shading is the *model-simulated* satellite picture for the same time, showing water vapor. This is an ECMWF twenty-four-hour forecast centered over the northeastern USA, starting at 12:00 noon February 6, 2011, valid at 12:00 noon February 7, 2011. There is a correlation between the dark and light areas of the image and the strong gradients in the PV field, shown by the contours crowding together. Comparing the actual satellite imagery with the latest forecast of that imagery, we estimate the accuracy of the forecast. Where discrepancies are apparent, knowledge of how the PV field relates to the wind, temperature, and pressure can be used to correct and improve the subsequent forecast. © ECMWF. Reprinted with permission.

Charney's 1948 paper, using the ideas of scale to identify balance, set us on the road to understanding how to capture the most important features of weather systems in terms of relatively simple mathematical models. Pixel versions of symplectic geometry in transformed variables could next enable those conservation laws to be built more effectively into computer models of weather.

EIGHT

Predicting in the Presence of Chaos

As the twentieth century came to a close, more aspects of weather and climate were being incorporated into computer programs. What is the state of progress with weather forecasting and climate prediction in the twenty-first century? Uncertainty does affect the way that we view the Earth's moisture-transporting atmosphere, and affects our ability to predict its changing behavior. Since we will never have perfect knowledge of the air and moisture above, we look at how to do better in the presence of the unknown, and how to improve computer representation of weather.

So we turn to the heart of the matter. Chapter 7 showed how, in principle, math allows the computer algorithms to respect the truths that lie behind the rules of atmospheric physics that were introduced in chapter 2. Now we need to cope more accurately with much more detail. In figure 8.1 we show the impact of a severe ice storm. The science of weather should be able to decide where and when this will happen, and determine the worst-case scenario. We describe how modern mathematics guides us to more realistic computer predictions of future weather.

The last two decades have seen supercomputer power and software development advance sufficiently so that the more delicate issues, which Charney had suggested in 1950 would concern only a certain few academics, might now be addressed more routinely. And predicting whether it will rain requires this. But first we describe a culture shift. Instead of trying to say exactly what the weather will be in the future, we instead try to predict what it will most probably be.

Figure 8.1. Freak weather can be spectacular and hazardous, but with climate change are we likely to experience extreme events more frequently in the future? To answer this type of question we need to understand how the "climate attractor" is changing. Once again, math helps us with this quest.

Prognosis and Probability

In 1964 Richard Feynman, having just been awarded the Nobel Prize for physics, gave a lecture to university students about general principles of science. In his inimitable way, he made one issue very clear: science is not always about proving absolute truths. As scientists, being able to say what is more or less likely does not necessarily compromise our integrity. To illustrate his point, he recounted a story of a discussion he had with "laymen" about the possible existence of unidentified flying objects. One individual kept pressing Feynman for an unequivocal "yes" or "no" as to whether he believed in the existence of UFOs. Finally, his antagonist demanded an answer—"can you *prove* they don't exist? If you can't, then you're not a very good scientist!" Feynman replied by pointing out that this view was just plain wrong. To make his point clear, he said, "From my knowledge of the world I see about me, I think it *more likely* that the reports of UFOs are due to the *known* irrational characteristics of terrestrial intelligence, rather than to the *unknown* rational efforts of extra-terrestrial intelligence."

More than a century ago, science was used to try to prove that the solar system would remain stable forever but found it could not, and a half century ago science could not prove that UFOs did not exist. One of our present-day concerns is whether science can give us definite answers about future weather and climate. Will chaos upset the applecart? Many popular remarks suggest it must. But if we look more closely at the toy model considered by Lorenz and the "butterfly wings" surfaces that he found in state space (shown in figure 7.7), we see another interpretation. Suppose that the key climate information that we seek—say, warmth or rainfall—is related to the shape of the butterfly wing. Then, even if we cannot predict precisely when, where, or for how long our life history will wind around on the butterfly wing, we may still estimate that average "climate" feature quite well. If the butterfly's wing changed its shape, then this would indicate a changing climate.

This approach requires us to focus more on typical life histories of weather rather than to identify the one true future that our weather will follow. Mathematically we refer to this as finding the weather attractor in state space. The word "attractor" is used to suggest that, even though individual weather events might appear random, weather events in general return to type, and the types have been studied with increasing care during the century since the Bergen School. Even at a more detailed level, the gusting wind over time periods of minutes to hours will maintain an average direction. Each time a new computer model is created to predict the weather, not only the way that we estimate or represent these averages but also the state space and the weather attractor inside the model change. But the weather itself should not depend on the way that it is represented.

Irrespective of the details of our model, the truths that should still hold and determine the shape of our weather attractor are the truths that help define the weather life histories for Earth. We need to ensure that each computer model has rules that sufficiently respect the conservation laws, those for mass, energy, moisture, and PV, for instance. Transformations in pixel variables, which focus computer algorithms on potential temperature and PV, help bring into view the conservation laws behind the rules, as discussed in chapter 7.

Further, as described in chapters 5 and 6, each weather future needs to be guided to appropriately respect the hydrostatic and geostrophic balance of the model's weather states so we can begin untying the Gordian knot of feedback. But how do we start a practical search for the attractor, either for tomorrow's weather prediction or for the prediction in several decades time? The simplest approach uses what has become known as *ensemble forecasting*. In the remainder of this section we describe how modern weather centers gain extra information about the reliability of forecasts through what is essentially a very simple technique.

Before describing ensemble forecasting in detail, we first explain how the notion of probability is used. As Feynman implied, probabilistic forecasting is a truly scientific endeavor. By a probability we mean the chance of something happening, often expressed as a percentage. Although we are often dismissive of information given in terms of likelihood or probabilities, consider how we might react to a weather forecast given in terms of the probability of heavy rain for a weekend when we are planning to hold a party.

The first scenario is a private garden party we are throwing for business clients who are about to make a decision on a major contract we have bid for. Because the stakes are high and we think it is imperative that our clients have a good time, then if a forecaster says that there is a 10 percent chance of heavy rain, we would probably consider that sufficient risk to spend around a thousand dollars to rent a tent. The second scenario is a similar party, but now the main guests are relatives and neighbors whom we want to impress. We consider renting a tent, but with a 10 percent chance of rain, we might save our money and run the risk of having to cram everyone indoors if it rains. The final scenario is the party to which we have invited all our buddies; we couldn't care less whether they get a little damp, and we certainly wouldn't entertain the idea of spending money on a tent. The moral of this tale is that if the loss–cost ratio is high, we'll often make decisions to take precautions even if the odds of a bad outcome are low; conversely, if the loss–cost ratio is low, then we might be prepared to take a gamble.

We contrast these scenarios with similar ones in which the forecaster does not offer information on the probability of rain but just issues a single deterministic forecast. If the forecasters say it won't rain and it does, then in the first scenario we would be justifiably angry. In the final scenario, we'll probably just have a good laugh at their expense. On a

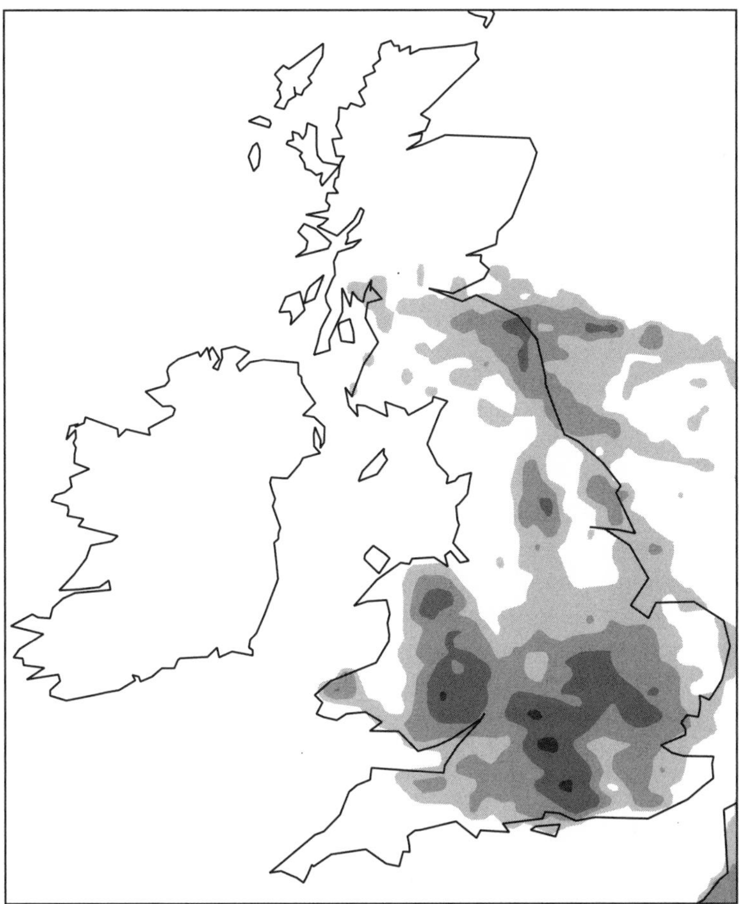

Figure 8.2. Here we show one forecast (made on April 26, 2011) of rainfall over the British Isles for the twenty-four-hour period Friday April 29 (midnight to midnight). Will it rain for the Royal wedding in London? The shading shows rainfall totals ranging from 2.5 to 25 mm. The way that we estimate the reliability of this forecast is to calculate "nearby" forecasts by calculating a series of forecasts starting from nearby initial states as shown in figure 8.3. © ECMWF. Reprinted with permission.

day-to-day basis, the vast majority of us deal with low cost/low loss decisions concerning weather, but there are many people in the world who have to make decisions in which the loss–cost ratio is huge: for example, launching a spacecraft, closing an airport and diverting the planes, or making a civil defense decision about whether to evacuate an area that might be hit by a hurricane.

Uncertainties will always exist because of the limitations of the models, because of errors in the data, and because of chaos, but the value of forecasts to decision makers is greatly enhanced if the inherent uncertainty can be quantified. This is particularly true of severe weather, which can cause such damage to property and loss of life that it is worth taking precautions even if the event is unlikely. Probabilities are a natural way of expressing uncertainty, and their use is increasing. By using probabilities, a range of possible outcomes can be described, and their likelihood can then be used to make informed decisions, allowing for particular costs and risks. Is it worth spending millions of dollars on flood defenses if the chance of a flood is less than 1 percent? Certainly not, if all that is flooded has very little value. However, in the case of New Orleans, the cost of the flood defenses and the value of the property defended runs into billions of dollars, so decisions have very serious consequences.

Many of us may feel cheated by not being told exactly what will happen instead being given what appears to be "vague" or "imprecise" information. But it is not really an issue of guesswork; it is an issue of how best to assess risk (or likelihood).

Forecasts can be most simply expressed as probabilities by making a list of the possible outcomes and seeing how often something that is of concern turns up. In figure 8.3, we show fifty-two forecasts for the heavy rain event of figure 8.2. Here the starting points equally represent our present knowledge—that is, the starting variations represent the unknowns or errors as fairly as possible. We then look at this ensemble of outcomes and count how many times it rains heavily at a particular location. A result of four times would be expressed as a probability of 8 percent that there will be heavy rain in seventy-two hours' time at the same location. This is ensemble forecasting: the process of repeating the forecast *N* times (*N* is the ensemble size) where we deliberately vary the quantities that we are least sure of to cover, or represent, most reasonably our ignorance of these unknowns.

Ensemble forecasting contains much more information than a single deterministic forecast, and this information is often difficult to convey to all users. Forecasts used on national television give a broad picture of the most likely outcome and warn of important risks. Each user's decisions may be based on the probabilities of a few specific occurrences. What these are as well as the probability thresholds for acting on the

Figure 8.3. We show fifty-two similar forecasts for the event of figure 8.2. These are not meant to be examined in detail. These forecasts are all generated by starting the same computer model from slightly different initial conditions. Forecasters estimate the uncertainty in their predictions by using "ensembles" of forecasts such as these and comparing the different outcomes. The common features in the predictions are usually forecast more reliably. If heavy rain is forecast in only four of these scenarios, then there is considerable uncertainty as to whether it will actually rain at that given location. The deterministic forecast here was on the pessimistic side; most forecasts were much more optimistic. © ECMWF. Reprinted with permission.

forecasts will differ. For important user decisions, it is necessary to apply each user's particular criteria to the detailed forecast information.

There are many ways of estimating the range of possible weather events. We could take a variety of plausible and usually nearby initial conditions. We could also vary the processes of cloud formation and the triggering of rainfall over the given time period of the forecast. Ensemble prediction is a practical method to find a single, deterministic forecast with an estimate of the probability of nearby forecast states. Ensemble forecasting provides forecasters with an objective way to predict the likelihood of each single deterministic forecast—in other words, to forecast the forecast skill. In practical terms, the flow of the governing equations is sampled, as explained in chapter 1 with the game of Poohsticks and illustrated in figure 1.19 for Hurricane Bill.

In figure 8.4 we show schematically how the information in the ensemble of forecasts shown in figure 8.3 might be represented graphically. What starts out as a neat compact set of nearby forecasts evolves ultimately into quite a range of possibilities, as shown by the final stretched and distorted image in figure 8.4. This is because of the effects of nonlinear feedback. The "true solution" and two possible deterministic forecasts for it are also shown. The spread of the possible nearby forecasts gives an indication of the reliability of the model in this instance (see figure 7.12).

In figure 8.5 we show the forecasts for air temperature in London given by thirty-three different forecasts started from very similar initial conditions for two different dates, June 26, 1995, and June 26, 1994. There is a clear difference in the way that forecasts evolve. In June 1995, as shown in figure 8.5a (top), the forecasts spread out uniformly and only gradually up to forecast day 10. In June 1994, as shown in figure 8.5b (bottom), all of the forecasts spread out rapidly from each other after forecast day 2. The way that nearby forecasts spread out as the days pass can be used as a measure of the predictability of the two final atmospheric states. Here we have a lot of confidence in the 1995 predictions and would confidently use a middle curve as the basis of the forecast to issue to the public. But those for 1994 suggest that something happens after two days so that we are not sure what will really happen next. Some of the forecasts suggest London's temperature will exceed 26°C after

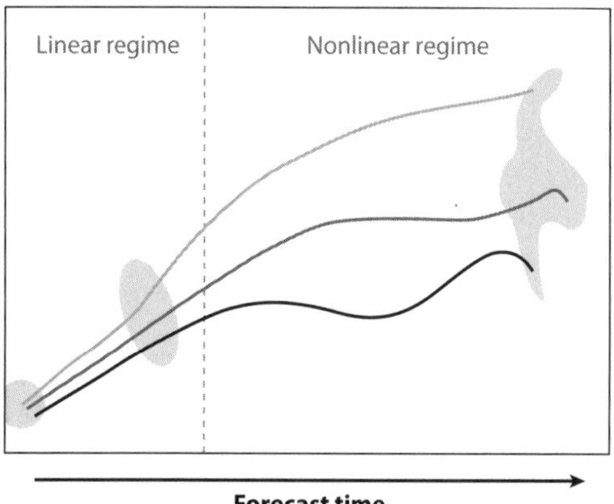

Figure 8.4. The deterministic approach to forecasting involves calculating a single forecast, shown by the lightest curve, for the "true" time evolution of the system, shown by the darkest curve. The ensemble approach to forecasting is based on estimates of the "probability density function" of forecast states, the evolving island shown initially and at two later times. We then estimate the reliability of the forecasts.

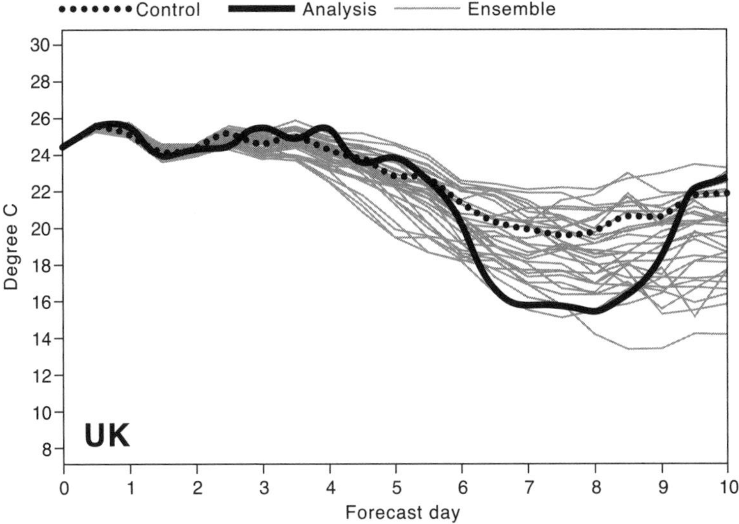

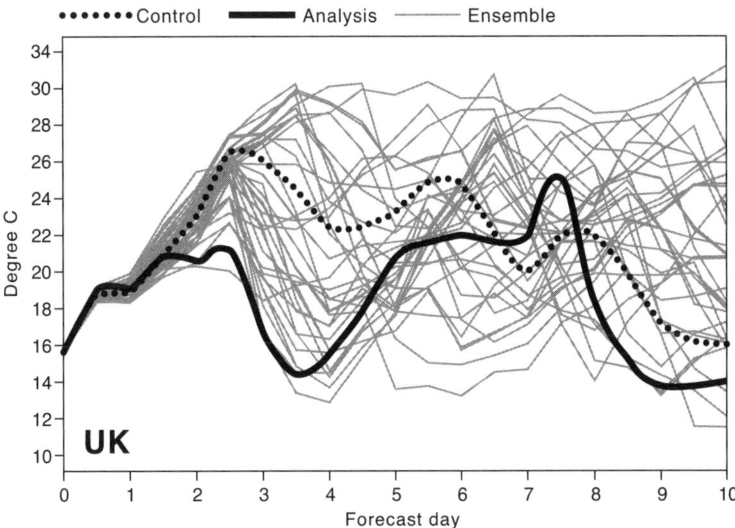

Figure 8.5 (a: top; b: bottom). The European Centre for Medium-Range Weather Forecasts (ECMWF), based in Reading, U.K., published forecasts for air temperature in London starting from (a) June 26, 1995, and (b) June 26, 1994. The amount of "spread" of the ensemble of forecasts after day 2 differs dramatically between the cases for 1995 and 1994. This evidence suggests that London's temperature in late June/early July was more predictable in 1995 than in 1994. Note that in case (b) the forecasting process is not very successful. In late June 1994, was the atmosphere in a state such that London's air temperature was essentially unpredictable? © ECMWF. Reprinted with permission.

day 6, while others suggest that the temperature will be below 16°C. The solid lines show what actually happened.

Ensemble forecasts are subject to the limitations of the numerical weather prediction models discussed earlier. Since very severe weather often impacts on relatively small regions, computer models may fail to reliably estimate the occurrence of these events. Together with the limited number of forecasts that can actually be run in any ensemble, this makes it difficult to estimate the probabilities of very severe or very rare events.

Over the seasonal time scale, and for most of the planet, forecasts of weather events have proved to be impossible. The hidden possibility of chaotic behavior of the atmosphere sets a fairly universal limit for present-day forecasting of the order of two weeks. This limit is often associated with the rapid growth of uncertainty originating in the initial conditions; these arise both from imperfect and incomplete observations, and from inaccuracies in the formulation of the model both in terms of its physical processes and its computational representation.

Weather pixel states are only known as accurately as the weather instruments can measure their values. Such small uncertainties will always characterize the initial or starting conditions for the computer calculation. Consequently, even if the weather pixel evolution rules were perfect, two initial states only slightly differing could sometimes depart one from the other very rapidly as time progresses, as Lorenz first showed in 1963, and as the computer predictions shown in figure 8.5b illustrate. The prediction of the early July 1994 summer temperature in London as anywhere between 16° and 26°C is an admission of failure for those computer models to forecast with confidence. Observational errors and errors in the details of the evolution rules usually appear first in the smaller scales of kilometers and minutes. Then they amplify and, through the feedback mechanisms, spread to larger scales, seriously affecting the skill of weather forecasts.

Following their introduction into general use in the 1990s, ensemble forecasting methods have remained a mainstay of major national forecasting agencies, with much study going into how best to "spread" the ensemble. This is done to estimate the reliability (in each region) of the daily forecast. By averaging over the ensemble, the forecasts that are most uncertain will tend to cancel out in the five- to fifteen-day prediction window. So the behavior of ensembles gives some insight into the nature of the weather attractor—that is, the nature of typical weather.

Diagnosis and Data Assimilation

In the previous section we described how modern forecasting uses a variety of starting conditions to predict the weather that will most likely happen next week. But there are millions of ways of equally realistically describing today's weather that are consistent with the observations. In this section we describe a technique for calculating better starting conditions.

The years following the introduction in the mid 1960s of practical computer forecasting have seen steady progress in the prediction process. Figure 8.6 shows improvements measured by the reduction of errors in the forecasts of pressure at sea level over the North Atlantic and Western Europe for the period 1986–2010. A quick glance tells us that five-day 120h forecasts in 2010 are better than the three-day 72h forecasts twenty-four years earlier. It is then a further challenge to improve the rainfall forecasts! Of course, this does not imply that every forecast is better than the previous forecast; it simply means that weather forecasts are, on average, getting better. Measuring forecast error meaningfully is a challenge—and often a public relations hot potato. Determining whether a forecast was any good depends on the type of information we require and the purpose for which we are going to use that information.

We note from figure 8.6 that the score for *persistence* (of pressure, in this example), which is a forecast that the weather for tomorrow is the same as today, has always been worse than the score for computer forecasts, and by a significant factor in spite of popular witticisms. Persistence is also an indication of how much the pressure at sea level changes over the period of a one-day forecast (and, although there is a natural variability, the values remain roughly the same). Therefore, the difference between this curve and the other curves is a measure of the skill of the forecast—the greater the difference, the greater the skill. Of course, comparing forecasts of sea-level pressure to assess improvements in the skill of forecasts is rather crude: such a measure tells us only a little about the skill of the models at predicting fog or ice.

So what do we mean by a poor forecast? If a model correctly predicts the time of the arrival of a severe storm but forecasts that the most intense part of the storm will move across an area eighty kilometers further north than its actual track, do we consider this to be a poor forecast? Given what we have said about the limited resolution of the

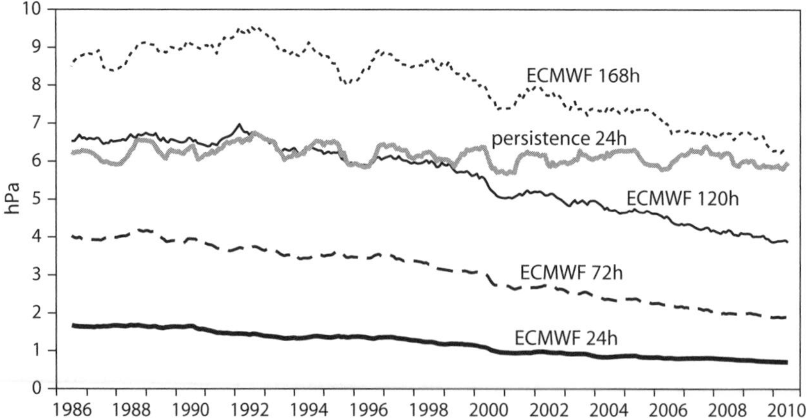

Figure 8.6. For the period 1986–2010, the improvement in forecasts is shown by comparing "actual" air pressure with the predicted air pressure at sea level. The vertical axis is the error in the predicted pressure in units of millibars (where 10 mb represents 1 percent of average sea-level pressure). These are the units that most frequently appear on a barometer. This highly average data indicates that forecasts are slowly but surely getting better. Persistence means that we use today's pressure as the forecast for tomorrow's. © ECMWF. Reprinted with permission.

models, this may not be a poor forecast for a global model with a large weather pixel representation. What if the rainfall prediction is correct in amount but arrives six hours later? This would not bother a farmer in his crops' growing season but might be a big deal for a major public event.

However we define poor forecasting, this illustrates that there is room for improvement, even taking into account the level of skill, as indicated in figure 8.6. Many customers, such as those planning to harvest grapes or delicate fruit and flowers, are especially sensitive at harvest time to forecasts of frost, local mist, and rain. The processes involving moisture are highly nonlinear with the trickiest feedbacks, and this often makes them the hardest to predict, especially at the local level.

The simple moral is that the more we want to forecast, the greater the challenge becomes. Forecasters are no longer simply asked to predict the large-scale distribution of sea-level pressure and the presence of weather fronts. Weather forecasting encompasses a wide spectrum of atmospheric motion, from the planetary scale, which affects the major weather systems and subcontinental heat waves and droughts, down to local storms and turbulence, which might lead to flash floods in valleys and towns or affect aircraft landings. Some flow patterns are readily altered and may be amplified using energy from moisture condensation and heating. These processes have strong feedbacks so that small uncertainties about the initial state of the local atmosphere may readily grow.

Eventually these sensitive patterns cannot be forecast. But how quickly does this happen? The details of thunderstorms cannot usually be predicted more than a few hours ahead—their development depends critically on sensitive but rapidly amplifying processes that also involve heat and moisture. On the other hand, the formation and track of hurricanes and cyclones can often be predicted up to a week ahead even when the rainfall prediction is less successful. The initial rates at which uncertainties grow is usually large—typically doubling in one to two days for larger-scale systems—so reducing the initial errors may only lead to a small improvement in the skill of a forecast, and a limit to accurate deterministic prediction probably always exists. But we certainly need to reduce initial errors.

When the basic equations of chapter 2 are used to calculate the changes to the temperature, winds, and rain over the next time interval, we need the temperature, winds, and rain values at the starting time, otherwise we cannot start. As Lorenz had shown, predictability is governed by (among other things) how, and how well, these initial conditions are estimated. Weather forecasters continually have to estimate what the most likely state of the atmosphere is given the actual instrument readings. They then base their next forecasts on this estimate, or *analysis*. In the remainder of this section we describe how a software process known as data assimilation is helping meteorologists replace unknowns in the forecasting procedure by best guesses using known information.

The problem of diagnosing the presence of a low-pressure system from measurements taken by barometer and thermometer is not actually that dissimilar to the discovery of the planet Neptune from indirect evidence, as mentioned in chapter 2: in both cases, a limited number of observations have to be matched, or fitted, to a conceptual model. In the case of a weather front, the model might be the Norwegian (Bergen School) model; in the case of a planet, the model is Newton's law of gravity. In both cases, partial observations are used to deduce "cause" from "effect."

There are many areas of science and engineering in which there is access only to indirect information about a process or a system. Examples include medicine and geological exploration. For example, a doctor who is attempting to diagnose an illness from information about a patient's condition, such as their temperature or blood pressure, or pictures obtained from scanning equipment, is presented with a set of readings or "observations," and the doctor's task is to work out what "cause" gives rise to these observations, which are the "effects." The challenge is to match the observations with the parameters or characteristics of the assumed model, thereby quantifying what is observed and giving us the potential to predict how things will evolve based on that model. Thus, a doctor might predict recovery under a certain drug regime. In geological exploration, shock waves are sent through the Earth's crust to create the causes, and returning signals caused by changes in the rock are measured (that is, the effects are measured). These echoes signal the presence of underground minerals or oils, and the job of the model is to predict where they are.

Mathematicians call the problem of deducing cause from effect an *inverse problem* because the task is to use "answers" to determine the cause, here usually described in terms of an equation. Up to now we have used equations to determine answers. The astronomers Adams and Le Verrier were engaged in a problem of precisely this type. In their case, the basic model was Newton's law of gravitation, but the mysterious anomalies in the orbit of the planet Uranus could not be explained on the basis of this law when applied to the then-known planets in the solar system. The missing ingredient in their model was the eighth planet, Neptune, and Neptune's presence not only generated extra equations but also modified the existing ones.

Before the advent of modern computer models, the conceptual models such as those developed by the Bergen School encapsulated our knowledge of how the weather variables would be related to each other in, say, a typical cyclone or low-pressure system. The problem for the forecaster was to fit that model to the scattered observations and draw the surface fronts on a synoptic chart. This process can be thought of as a subjective inverse method. The human mind is very good at second-guessing what a "near-best answer" should be, but we need to automate the process and exploit the ever-growing power of computers to get more precise or accurate estimates than humans are capable of. The computer can also do the calculations at all necessary places around the globe in a fraction of the time that humans could.

This type of problem has a long history, and once again the challenges of astronomy spurred many of the key mathematical developments. At the turn of the nineteenth century, mathematicians such as Karl Friedrich Gauss and Adrien-Marie Legendre were preoccupied with the problem of determining the orbits of comets that astronomers were discovering with the latest powerful telescopes. Gauss realized that the observational information was imperfect because errors would inevitably be made in taking the observations of the positions and motion of the objects. He also realized that the laws of gravitation used to determine the orbits were also incomplete in the sense that we would never be able to account for all the small effects that would actually influence the motion of a comet (for example, the gravitational field of another distant planet or the presence of asteroids). He therefore concluded that the process of determining the orbits by using a mathematical model and observational data was intrinsically inexact. The question was how to deal with such a problem.

In 1806, in a paper entitled "Nouvelles méthodes pour la détermination des orbites des comètes" ("New Methods for Determining the Orbits of Comets"), Legendre set out his equations in a way that would account for the inherent inexactness of the problem. So, for example, instead of writing $a + b = d$ in which, if a and b are known, d can be computed exactly, he wrote $a + b = d + e$ where e denotes an error representing our "not quite right" values for a and b. Legendre had many equations to solve, and in writing them in the form described above, he then hypothesized that the best solution would be the one that minimized the sum of the squares, $e \times e = e^2$, of the errors. This method became known as *the method of least squares*, and it is the basis of many modern techniques for finding the best analysis. Minimizing the sum of the squares of the errors in this way is equivalent to maximizing the probability distribution. That is, *minimum error equals maximum likelihood* of the data being observed subject to the known constraints controlling the values. This idea of finding a "best" straight line is given by a simple illustration in tech box 8.1.

In the data assimilation methods as used in present-day weather forecasting, we look for the "best forecast" by comparing forecasts with observations over a restricted time period, often the next six hours. This best forecast is then used to give the pixel starting values for the computer forecast over the next few days. There are always insufficient

Tech Box 8.1. Determining Causes from Effects

A very simple example of how "answers" (or effects) can be used to find a particular form of an equation (here, the cause) is illustrated by figure 8.7. The straight line represents a linear relationship between two variables, x and y. The key *parameters* needed to determine a specific straight line are its slope, m, and the point, c, where it meets the y axis. Suppose there are observations of how y varies with x—these are represented by dots in the figure. We happen to know that the process or system we are trying to understand is described by a linear model; what we do not know are m and c.

So the task in an inverse problem is to use the observations, the dots that represent the information we have of y and x, to determine the slope, m, and the intercept, c—in other words, to place the line as well as possible through the data. Note that there is no perfect answer—either the solid or the dashed line shown in figure 8.7 could be correct. The best answer depends on how we value the various errors. In a perfect world, the model straight line would pass through all the data points. We interpret the fact that actual data do not lie on the line to errors in the measuring and observing process, or to other deficiencies not yet understood. It is even possible that we erred in thinking that the model is a straight line.

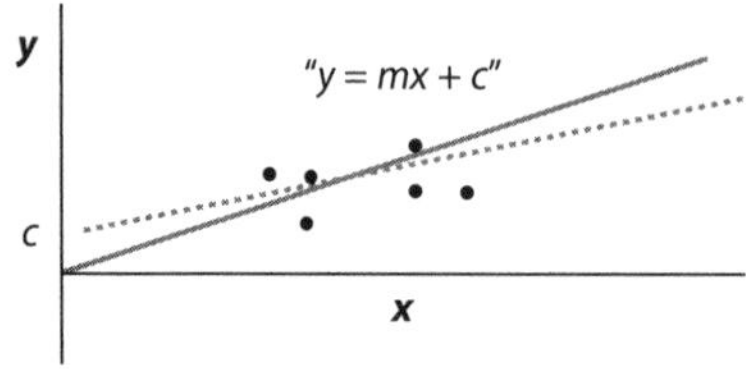

Figure 8.7. Which straight line best fits the data?

observations at any one time to determine the state of the atmosphere on all of the pixels. If we want a more detailed picture, then additional information is necessary. Data assimilation is therefore the process of finding the model representation of the atmosphere at the starting time that is most consistent with a limited set of observations.

Forecasts today routinely start from a description of the state of the atmosphere that has been built up from past and current observations.

This data assimilation procedure is designed to help with both the inadequacies in the physical process representations and the errors in the initial charts, as we have discussed in the final section of chapter 6 and in tech box 8.1. The assimilation process uses the computer weather prediction model to encapsulate and carry forward in time information from previous observations and forecasts. Data assimilation is very effective at using the incomplete coverage of observations from various sources to build a coherent estimate of the atmospheric state. But, like the forecast, it relies on the model and cannot easily use observations of scales and processes not represented by the model. This means that a great deal of information used to start a forecast has the status of "best guess": there are many unknowns and, as we shall explain later, even some "unknowables." These limitations in current models particularly affect detailed forecasts of local weather elements such as cloud and fog. They also contribute to uncertainties that can grow rapidly and that ultimately limit predictability. But what is the best starting state for a forecast? Sometimes starting with deliberate "errors," where the computation is guided to respect the pixel weather attractor, helps us to forecast even better.

The initial state of the atmosphere is very difficult to determine despite the fact that millions of observations of the atmosphere and oceans are made from land, sea, and space every day. In fact, the weather forecaster continually has to contend with the limitations imposed by the *lack* of observations of the current state of the atmosphere. Why? Well, despite the fact that we have access to observations from many different locations on the ground, from within the atmosphere, and now from above it using expanding networks of satellites, there are still huge data-sparse areas where there is little or, at best, very indirect knowledge of the state of the atmosphere. Although new satellite observing processes should give a better idea of quantities such as the moisture distribution, the detectors cannot "see" through all clouds, and often it is just beneath the clouds where the interesting weather is taking place. Consequently, there are many gaps in the horizontal and vertical coverage that we have to fill before a new forecast can be made. See figure 8.8, where the data on ozone obtained by a satellite making several passes around the planet has been smoothly filled-in. Furthermore, the measurements themselves will not be precise.

One of the main sources of information for a new forecast is, in fact, the previous forecast. The previous forecast helps fill the data voids, and

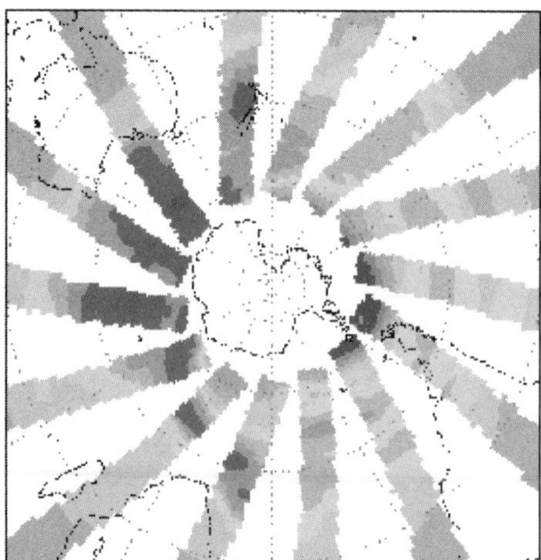

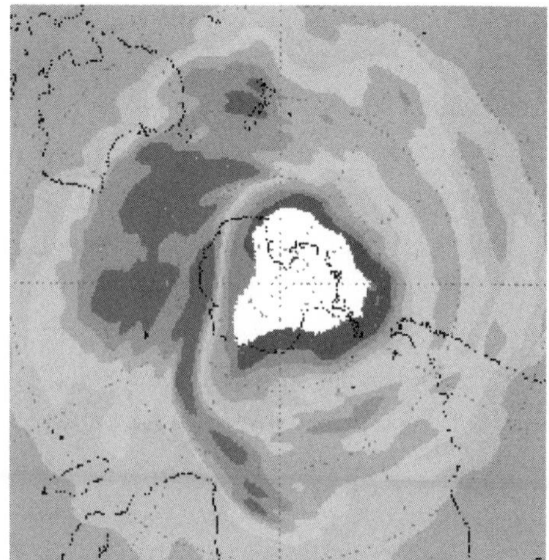

Figure 8.8. On the left is an image of the southern hemisphere depicting what a satellite "sees" below it while orbiting the Earth and gathering data (which might be temperature, water vapor, or ozone, for example, as here). On the right is the product of data assimilation, which provides the missing pieces of the three-dimensional jigsaw (remember that the atmosphere has depth). This is how the information is incorporated into the computer models—a potential source of error and computationally very costly. Courtesy Alan O'Neill.

the task of making a new forecast begins with incorporating the latest observational data with old, previously calculated data. For example, if we make one forecast out to six days ahead every twelve hours, then the first twelve hours of every forecast can be updated with new data obtained during that time to produce the best-improved initial conditions for the next forecast. Using the previous forecast means that we might inherit any errors present in that forecast, and this is another one of the unknowns in the forecasting game. This is why the process is called data assimilation.

New data has to be incorporated in such a way that it is consistent with the previous forecast while allowing for errors in both the data and the old forecast. Because we make measurements of different variables, the assimilation also has to make sure that the relationships between the variables are consistent with the laws of physics (which the errors in the observations may not be). Even accurate observations at a point (totally consistent with the physical laws at this point), if used over a pixel, can still lead to contradictions with these basic laws. For instance, the assumption that temperature is constant over the pixel may not be consistent with the way that the change in pressure and density was calculated in that pixel. The whole process is extremely complex and computationally exceptionally demanding. At the conclusion of the assimilation process, it is vital to make sure that the initial conditions are as consistently accurate as possible: Richardson did not specify his initial conditions correctly, and this led to his catastrophic error in the pressure forecast after only six hours.

Better and more numerous observations have been a significant part of the improvements to computer simulations that have contributed to the successful advance in weather prediction over the past forty years. The sources of data used by the forecasters are many and various. They include ships, aircraft, oilrigs, buoys, balloons, manned land stations, and satellites. Automation has to a large extent replaced the human observer, and use of this technology provides information from remote, inhospitable, or inaccessible places.

New generations of satellites gather more, and more accurate, readings, ranging from the sea surface temperature to the state of the stratosphere. Data is exchanged freely around the world among weather bureaus; global weather prediction relies upon this protocol. However, we will never have perfect, complete weather data. The introduction of ever-more pixels to resolve ever-finer detail means that less than 1 percent of the necessary data is being directly observed at present. Data assimilation has to fill in the missing 99 percent plus. The patchiness of rainfall, as illustrated in figure 8.9, is a particular difficulty. Improving data assimilation software allows forecasters both to get the best value out of known data, and to improve the software that models physical processes.

Present-day weather forecasting on supercomputers must deal with imperfect knowledge of what the weather exactly is today and how physical processes such as cloud formation precisely work. In use since the 1990s, modern data assimilation software addresses these issues, although much effort is still devoted to improving the algorithms. But the techniques of ensemble forecasting and data assimilation have to contend with an issue that Charney put to one side—the issue of the "overtones" in the symphony of atmospheric motion. This issue is brought to the fore as the size of the computer pixels decreases and model resolution increases—although the detail is important, we still have to be able to see "the forest for the trees."

Figure 8.9. Cumulonimbus capillatus, photographed off Falmouth, Cornwall, England, on August 12, 2004, show a sharp boundary or discontinuity that separates where it is raining from where it is dry. This cell developed quickly over the coastal fringes of southwest England, giving a short heavy thunderstorm and a very marked squall inland, while less than 5 km off shore the sunshine remained unbroken. Image © Stephen Burt.

Listening to the Overtones

The continually improving capacity of supercomputers to carry out billions of calculations in seconds gives modern forecasting agencies the ability to describe much more weather detail, and to warn us in good time about dangerous weather. This has required the pixel rules to be drafted extremely carefully, so that much more mathematical detail in the original equations of chapter 2 is captured. We first describe the types of weather that need to be focused on and then discuss the ongoing refinements to these pixel rules.

The dynamic and dramatic nature of the atmosphere seems to demand our attention and understanding more than ever before. Were the floods of August 2002 in Europe and Hurricane Katrina in August 2005 in the Gulf of Mexico unfortunate but rare events? Both China and

Pakistan have experienced heavy monsoon rains and extensive flooding during July and August 2010, which seriously affected millions of lives. The amount of resources that should be focused on building better flood defenses depends on how often such events will recur and what form they will take. In the meantime, flood insurance for such events becomes a cost for us all in both the price of goods and government taxes.

In many countries, government agencies have been tasked to develop programs that predict various hazardous phenomena such as damaging winds that spread pollutants and fan bush fires, the paths of tropical storms and storm surges, heavy rainfall and consequent floods, and drifting volcanic ash clouds. More localized models provide the forecaster with detailed information that helps environmental agencies take action in specific locations to minimize the threat to people, property, and livelihood. Models of the atmosphere–ocean system that are more complete and are linked with land-based ecologies and the frozen regions lead to climate simulations that help governments decide what actions could be taken to adapt to the effects of climate change. Over the past two decades, increasing the resolution of the pixels, designing better ensembles, and finding more efficient ways of exploiting limited data have all led to more detailed and more reliable forecasts. By 2005 many global forecasting centers were using pixels that had an approximate horizontal dimension of twenty-five kilometers, but this is often not detailed enough to describe the natural disaster. So these hazard-warning agencies focus on limited-area models where pixel dimension for 2010 is closer to one kilometer in the horizontal in order to resolve the serious flood, fire, or tempest in specific localities.

Hazardous weather has a major impact on health and safety and on the economy in most countries. The cost of the floods in autumn 2000 in the United Kingdom alone has been estimated at £1billion, and a staggering $80 billion of total damage followed Hurricane Katrina. Insurance companies worldwide now pay out more than $4 billion annually in respect of specific flood damage. Flood events are often localized and associated with small-scale meteorological phenomena, in particular, thunderstorms with their brief deluges. Local storms can give rise to the most dangerous floods, to hail and lightning, and to extreme local winds, including tornadoes. In 2010 supercomputer-based weather prediction models still did not have sufficiently fine pixels to adequately represent all of these hazards.

However, these studies have demonstrated that computer weather-prediction models that run at very fine resolutions of one kilometer or less are capable of simulating the evolution of individual storms. The development of computer power enables models where an entirely new pixel world occupies one old weather pixel from the previous generations of computer models. The effective solution of this fine-scale weather embedded in the large-scale models and the relation of the fine-scale weather to the original grid are crucial ongoing issues. In figure 8.10 we show schematically the weather features that various length and time scales can resolve. The larger the computer model, the smaller the pixel size, and so the more detailed the weather phenomena that the computer model can resolve in principle, and mostly in practice.

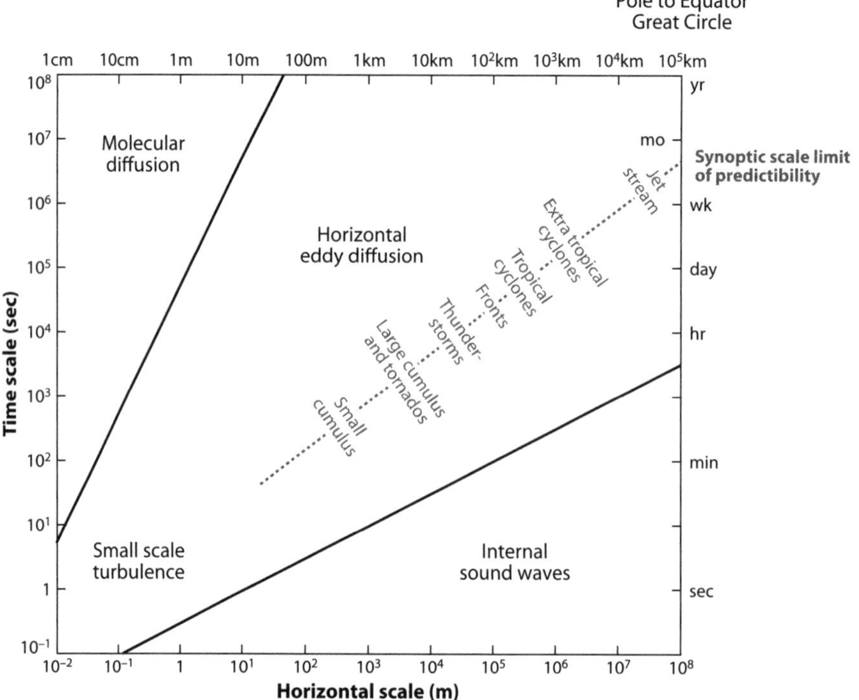

Figure 8.10. Familiar atmospheric phenomena ranging from small cumulus clouds through tropical cyclones (hurricanes) to the motion of the main jet streams are plotted against their typical length and time scales. The aim of more powerful computation is to have smaller weather pixels that effectively represent smaller weather phenomena. At present, local forecasting models use pixels that are about the size of large cumulus clouds, while global models have pixels with horizontal extents of more than twenty kilometers.

Keeping track of hazardous weather requires the use of radar and satellites together with information from the computer models. Straightforward extrapolation of this information into the future is a valuable technique (which harks back to the days of FitzRoy and Le Verrier), providing warning of heavy precipitation, wind, and fog for several hours ahead. Other hazardous phenomena, such as hail and lightning, can be predicted for up to an hour or so ahead. Such "nowcasting" schemes, as they are called, based on extrapolation rather than on numerical weather prediction, currently provide the best source of guidance for the issue of warnings to industry and to the public. In the short term, gains in predictability will follow improvements to nowcasting schemes and their data input. Bigger gains will come from exploitation of the new generation of satellites and computers, which will provide even more detailed information.

At the larger, more countrywide horizontal dimensions or scales, some of the processes occurring in the atmosphere are coupled with those in the ocean, and with the land surface. Attempts are now being made to predict the response of the atmosphere to a variety of forcing mechanisms, such as the motion of warm- or cold-water currents, the growth of crops, and the drying out of swampy land. These processes occur over months, and although they have inherently longer time scales than most atmospheric motions, there is still a significant linkage, which is important for climate prediction.

Crucially, weather systems have to be several times the size of the weather pixels to be represented accurately (recall figure 6.11). For example, the formation and evolution of low-pressure cyclones and high-pressure anticyclones has been simulated more accurately post-2000 because these phenomena have dimensions that are typically at least five or six times greater than the size of the computational pixels used in global models. Events such as individual convective cloud development that occur on scales much smaller than the pixels and yet influence the larger-scale weather have to be represented by averaging to the scale of the pixels.

As an example of important physical detail that can never be represented on weather pixels, we mention the wide variety of water droplets, snowflakes, and ice particles that occur in clouds; the average effect of this precipitation is estimated using a mixture of physical and computer experiments, as are the effects of the clouds themselves and the underlying forests, farmland, and urban areas within each pixel. The sea surface, its waves, and evaporation are also only represented in an average sense. With pixels sometimes covering areas of much more than one hundred square kilometers, it is clear that many aspects of "local" weather—from dust devils to isolated showers and wind gusts—cannot be represented. Even with the undoubtedly staggering improvements in supercomputer technology in the coming decades, the resolution of the models will still not be able to represent *all* atmospheric motions and processes. Individual gusts and the swirling edges of clouds might remain uncomputable for many decades to come. We show a typical example of the complexity of the evaporation and precipitation processes of such in figure CI.14. Our point is, no matter how big supercomputers become over the coming decades, forecasting agencies will still not be able to resolve all the physical detail (see figure 8.11). For the foreseeable future, the issue remains of how best to predict when much detail remains unknowable.

We next look carefully at the weather rules of chapter 2 in order to identify additional mathematical truths hidden in those equations. One

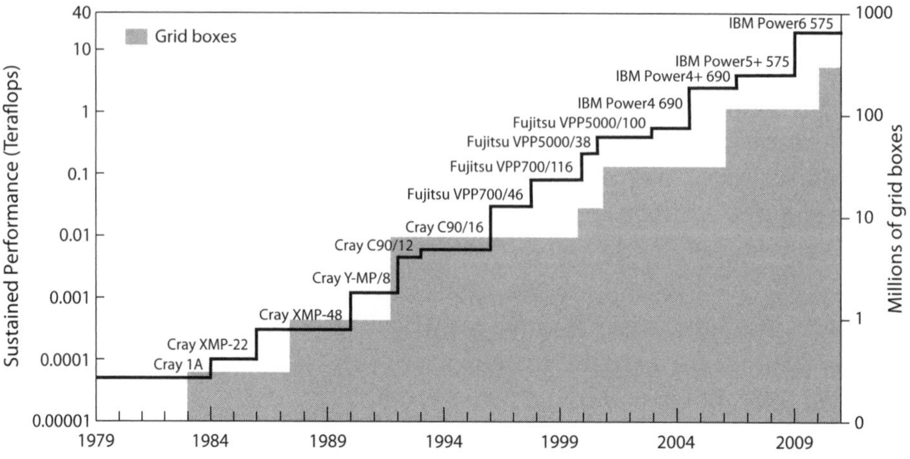

Figure 8.11. To improve the resolution of the model by a factor of 6 in terms of horizontal resolution (which means that a weather pixel with an east–west and north–south dimension of, say, 12 km is replaced by thirty-six 2 km by 2 km base squares) requires 140 times more computer power. By Moore's Law (which says that computer power doubles about every eighteen months), this takes about a decade. But improving computer power and capacity brings further challenges. The approximations involved in modeling detailed heat and moisture processes need to be significantly improved at much finer pixel resolutions. © ECMWF. Reprinted with permission.

truth that now needs to be respected is associated with what we called in chapter 6 meteorological noise, which Charney said only concerned the academics at MIT and NYU. Although we saw in chapter 6 that much detail can be ignored when making the daily weather forecast, we need to be more careful when making higher resolution predictions.

Today computation of the pressure, for instance, needs to be at much better than the one-millibar level. This is approximately one hundred times more accurate than the simple calculations for the next day that ENIAC carried out. Although a millibar is only 0.1 percent of the atmospheric pressure at sea level, it is still about 10 percent of the dynamic horizontal pressure gradient force that drives the ever-changing winds. This obliges us to pay the overtones of Charney's musical analogy their due respect. The effects of some of the pressure oscillations are visible in figures CI.8 and 6.4 where patterns in various cloud layers are identified as gravity waves. The fundamental assumption made in chapter 6 about the use of perturbation theory to cut the Gordian knot of nonlinear feedback suppressed these fast gravity waves of influence in the atmosphere to achieve the state of geostrophic balance. But as we continue the perturbation approach, thus linking cause and effect more accurately, the effects that were initially ignored are brought back into view, and eventually the gravity wave effects must be balanced appropriately.

Modern computer programs radically changed in the decades of the 1990s and 2000s in order to deal more effectively with all of these issues. The evolving supercomputer power enabled effective modeling of the gravity and fast-wave detail combined with the conservation of air mass, energy, and moisture. The basic strategy has always been to use software schemes that compute from now to, say, a few minutes from now, the "time step" in chapter 3—although Richardson's time step was several hours. This time step needs to be sufficiently accurate so that the software may be used again and again many thousands of times to advance the forecast to many days ahead.

The present schemes with which national forecasting agencies compute amount to contemporary versions of Richardson's model, which used finite difference representations of the fundamental physical rules written in their Eulerian form, as described in chapter 3. There are three main ways that this process has been improved. The first is to use implicit representations of certain key time derivatives. These are referred to as semi-implicit (SI) methods, and they limit the destabilizing effects of the fast waves. The second is to respect the so-called Lagrangian formulation of the *D*/*Dt* operator on pixels, as used in tech box 2.1 in chapter 2; these are referred to as semi-Lagrangian (SL) methods. The final improvement has been to modify such semi-implicit/semi-Lagrangian (SI/SL) methods so that they sufficiently accurately conserve the air mass, energy, and moisture as these quantities are being transported by the airflow. The prefix "semi" indicates that judicious choices are made when formulating the pixel rules. The computer algorithms that implement these procedures should also simultaneously satisfy the hydrostatic, geostrophic, and PV perturbation results of chapter 6, with special modifications in the tropics. Since no one has yet found a way to do this exactly, research continues.

Applying these three major ideas leads to algebraic equations for the pixel quantities that, in general, cannot be solved sufficiently quickly and accurately to allow the tens of thousands of time-step evaluations necessary for a weekly forecast. So the ideas of hydrostatic and geostrophic balance (discussed in chapters 2, 5, and 6) are combined with the more accurately conserving SI/SL methods. Then clever computer algebra is exploited to iterate the solution procedure and to enable a speedy solution that is hopefully sufficiently accurate. When the solution procedure on the latest supercomputers is fast enough, a more accurate weekly forecast, together with its ensemble, becomes possible. If the procedure remains stable, which means that the butterfly effect of error growth remains controlled, then medium range forecasts become possible. A computer calculation involving a million time steps even becomes a practical possibility, and climate forecasts then ensue.

We now go into a little more detail about the key ideas behind these methods. In chapter 4 we described how temperature in a balloon changed as both time went by and the balloon drifted along with the wind. We have already thought of an air parcel being blown or transported by the wind in a way similar to the motion of the balloon. Consider how to evaluate the change of, say, the temperature of this air parcel. To work out the change in parcel air temperature in a Lagrangian manner over the time step, it is necessary to find out where each parcel moves across the grid of pixels.

Suppose we focus on an air parcel that occupies a weather pixel at the time step of one minute ahead. Then we have to work out where this air parcel is now in order to know its starting temperature. Usually this air parcel will no longer correspond to any given pixel but will involve a collection of pixels. Calculating the present position of the air parcel thus involves calculating the position of a "control volume," which we denote as CV. In order not to gain any spurious air over the time step, the mass of air should be exactly conserved as the air parcel moves from the CV to the new pixel location. But we do not want to gain any spurious heat energy either, so this should also be conserved, as should the moisture being transported by the air parcel. At present, no way is known for calculating these quantities, so that they are all *exactly* conserved.

There is a further difficulty if a weather center decides to use massively parallel computer architecture, when ten thousand, or perhaps as the decade advances even one hundred thousand, processors compute simultaneously, each on their part of the weather pixel description. Rather than use the CV idea, a more accurate representation of the *D*/*Dt* operator on pixels may turn out to be more effective. As usual, each weather center chooses its own way to balance the requirements of accuracy, stability, and efficiency as well as its own way to model the physical detail.

In fact, the faster-moving signals such as gravity waves keep the atmosphere moving in harmony with the undulations of the terrain underneath, with the convective storms of the tropics and the extra tropics, and with the pulsations of the stratosphere above. Even so, harmony occasionally breaks down. Every now and then, for example, large chunks of the stratosphere are "folded" in to the troposphere, mainly associated with the movement of the jet stream, which has significant consequences for the weather. If we do not correctly sum up all these minor gravity wave effects to begin with, we usually do not successfully predict the major outcomes some time later.

Untangling the feedbacks as the wind blew the warm and moist air parcels, which were themselves interacting in ways that steered the winds, made the first successful computations possible. The untangling relied on Charney's basic state of the middle latitudes to perturb about, the weather state that respected hydrostatics in the vertical as well as geostrophic balance for the winds and pressure field in the horizontal. Charney's basic state also respected the typical variation of temperature with height that represented the local climate. Today better observations and data assimilation get these typical weather balances into the starting states, and the time-step evolution rules lead to an overall balance being maintained as the forecast proceeds. Forecasting centers are now focusing on how to deal with convective instabilities to improve the precipitation parts of this process, and they are finding challenges in the part of the code that deals with the tropics.

Certain large-scale stable structures in the atmosphere are visible from space—structures such as organized clouds of tropical hurricanes and middle latitude cyclones, which we see in figures 3.8 and CI.7. These swirls, jet streams, and fronts are usually major influences on the accurate simulation of local weather systems, and present-day computer models do a reasonable job with this. Since, to a good approximation, the evolving large-scale wind patterns can be described by conserved quantities, the overall quality of weather forecasts in nonstorm situations is much better now than twenty years ago. However, for deep tropical convection and storm-scale forecasting, it is vital to maintain the near-unstable conditions in the computer simulation while at the same time correctly triggering updrafts and latent heat release. This demands the accurate representation of rapidly changing structures at scales of less than one kilometer that are still below the pixel scales of the usual weather forecasting models. How can we conserve thermal energy and moisture in such circumstances? We return to this issue in the final section. But first we discuss in the following section how very lengthy calculations are made to estimate the changing climate.

From Seconds to Centuries

Climate models involve fundamentally the same computer models that are used for weather forecasting for periods up to a week ahead, but they are linked to models of many other components of the Earth system, such as the oceans, the ice caps, and the planet's albedo (see figure CI.15). Typically the climate models are run at a coarser pixel resolution and for decades instead of days. To compute twenty, fifty, and more years into the future, much finer-scale detail has to be ignored. It is hoped that successful prediction of major changes such as the development of areas of semipermanent flooding or aridity on countrywide portions of continents remains possible. The larger pixel sizes used, for instance, usually means that the effect of the Andes mountain range on eastward winds and rainfall over the Brazilian rain forest is not computed as effectively. It is not known to what extent this undermines the accuracy of present-day long-time global predictions of climate. However, even though the detail is not being predicted, it would appear necessary to compute even more carefully the changing temperature and moisture balances across the pixels. Since the calculations have to run for decades, a small loss of heat or moisture that starts out as a negligible error over a few weeks might actually grow to a size that masks the real changes calculated fifty years ahead.

Weather and climate have a profound influence on life on Earth. The weather is the fluctuating state of the atmosphere around us, whereas the climate is the "average weather" and its slow evolution over the decades. More rigorously, it is a statistical description of weather, including variability and extremes as well as averages; climate also involves the land, the oceans, the biosphere, and the cryosphere (see figure CI.15).

To make predictions about climate change, the effects of all the important links operating in the climate system have to be calculated. Knowledge of these processes is represented, following Bjerknes's original manifesto, in terms of mathematics and is implemented today in a computer program which is referred to as a climate model. The overall rules of the atmosphere module are those of chapter 2, but the cloud, rain, and heating processes are now being considered more carefully. The ocean module follows similar rules, with the salinity taking over and modifying the role of moisture. These two modules are coupled by the moisture exchange at the sea surface, and wave generation significantly influences this. Modules for the other climate processes are then added, and their feedbacks on each other further complicate the computer output.

The limitations of our knowledge and computing resources mean that the results of climate models are always subject to many uncertainties, but this does not stop us from proceeding in ways that Feynman would have approved. Atmospheric circulation—the ceaseless winds that transport the warmed and moistened air around planet Earth—is fundamental. In climate simulations, processes such as the cycles of trace gases, including methane, sulfur dioxide, carbon dioxide, and ozone as well as their chemical interactions, together with dust (see figure 8.12) are also represented. The long-term change of these atmospheric constituents usually accompanies major climate change, and part of present-day science tries to understand, for example, how oceanic plankton blooms and loss of tropical rainforests modify these gas concentrations.

Figure 8.12. Saharan dust, rich in nitrogen, iron, and phosphorus, helps to fertilize the huge plankton blooms that occur in the tropical eastern Atlantic and Mediterranean. The picture shows dust being blown in northward and westward directions. © NEODAAS / University of Dundee.

The nature of predictability in climate may also be understood in probabilistic terms. It is not the exact sequence of weather that has predictability in the years ahead but rather aspects of the statistics of the weather—for example, in twenty years' time, will summers be warmer and wetter? Of course, the answer depends very much on where we live. Though the weather on any one day may be entirely uncertain so far into the future, the persistent influence of slowly evolving sea surface temperatures may change the odds for a particular type of weather occurring on that day.

In a rough analogy to the process of throwing dice, the subtle effects of the coupling between the atmosphere and its surroundings can be likened to "loading" the dice. On any given throw, we cannot predict the outcome; yet after many throws, the biased dice will favor one particular outcome over others. By following this analogy, changes in seasonal behavior years ahead may still be predicted even though we cannot predict the actual weather at a given place on any given day. On these very long time scales of decades, the predictable part of climate is the structure of the weather attractor that these computed life histories eventually remain near or keep visiting. A major part of estimating climate change lies in understanding what significantly modifies these "butterflies' wings" as the decades roll on.

The ocean is a crucial part of the climate system. There is a constant exchange of heat, momentum, and water vapor between the oceans and the atmosphere. Even oceanic salt particles in the atmosphere are considered to have significant effects on water droplet formation. The ocean acts as a heat reservoir that initially delays climate change. The major ocean currents transport large amounts of heat energy, together with seawater of varying salinity, around the world. The land surface, including its vegetation and seasonal moisture distribution, affects the absorption of the Sun's energy as well as the cycling of moisture with the atmosphere.

Of note also is the cryosphere, especially those parts of the world whose surface is affected by ice, principally sea ice in the Arctic and Southern Oceans and the land-based ice sheets of Greenland and Antarctica. The biosphere, which includes life on land (the terrestrial biosphere comprising forests, grasslands, urban communities, and so on) and in the ocean (the marine biosphere comprising seaweed, bacteria, plankton, and fish life), plays a major role in the water vapor, oxygen, sulfur, carbon, and methane cycles, and, hence, in determining their atmospheric concentrations. These trace gases in the air then affect the way that the solar energy transfers through our atmosphere and remains in it. The point in mentioning all these systems is to indicate that their many nonlinear feedbacks become a major complication in reliably estimating temperature, moisture, and climate in the twenty-first century.

The climate is always changing as planet Earth moves from ice age through warmer times and back to ice age. Feedbacks in the climate system can act either to enhance or to reduce these changes. For example, as the atmosphere warms, it will be able to "hold" more water vapor. Water vapor itself is a very powerful greenhouse gas, as we can readily tell when cloud cover at night slows the Earth's cooling rate. Water vapor could act as a positive feedback and significantly increase the average atmospheric temperature. When sea ice begins to melt, some of the solar radiation that would otherwise be reflected from the sea ice is absorbed by the ocean and heats the surface water layer further—another positive feedback. But we still need to find how these "warmings" will then modify the global wind patterns, jet streams, and ocean currents, and find what the next feedbacks will be. On the other hand, carbon dioxide concentration increase in the atmosphere acts to speed up the growth of plants, trees, and plankton (which is known as the fertilization effect), and this in turn absorbs more of the carbon dioxide; this is a negative feedback. There are multitudes of feedbacks, both positive and negative, many of which we do not fully understand. Feedback is the main cause of uncertainty in climate predictions, and changes in water vapor and cloud cover are crucial to accurate predictions.

Present climate models give a range of predictions in global-mean quantities; the uncertainty at smaller scales is bigger. Because we have no way of assigning the skill of each of the models, all of the predictions are often assumed at present to have the same (unknown) chance of being right. This is obviously unhelpful to planners trying to formulate new lifestyles. For example, hydrologists have to decide whether a new reservoir should be built to avoid the predicted summer water shortages. It costs resources to build large reservoirs, so what should the smallest safe provision be? Damaging changes need to be evaluated, effective action identified, and priorities established. Even more importantly, to what extent is controlling carbon emission to the atmosphere the best use of present resources? Would other actions be more effective?

The ozone hole over Antarctica has been studied extensively by computer simulations. The loss of ozone caused by chemical pollutants released mainly in the 1950s and '60s has been halted, and the ozone hole is now healing. The international cooperation on banning chlorofluorocarbons from household items is allowing the ozone to re-form—a big success for concerted governmental efforts.

The increase in accuracy and certainty of supercomputer climate predictions is unlikely to be rapid, depending as it does on hard-won improvements in understanding how the climate system works. To decide on the optimum planning and adaptation strategy, planners want to move away from the current situation of having a large number of different predictions of unknown credibility and move to a situation where the probability of different outcomes (for example, percentage changes in summer rainfall) is known. Recognizing that models give different predictions because they use different representations of the climate system, a varied selection of climate models have been constructed and

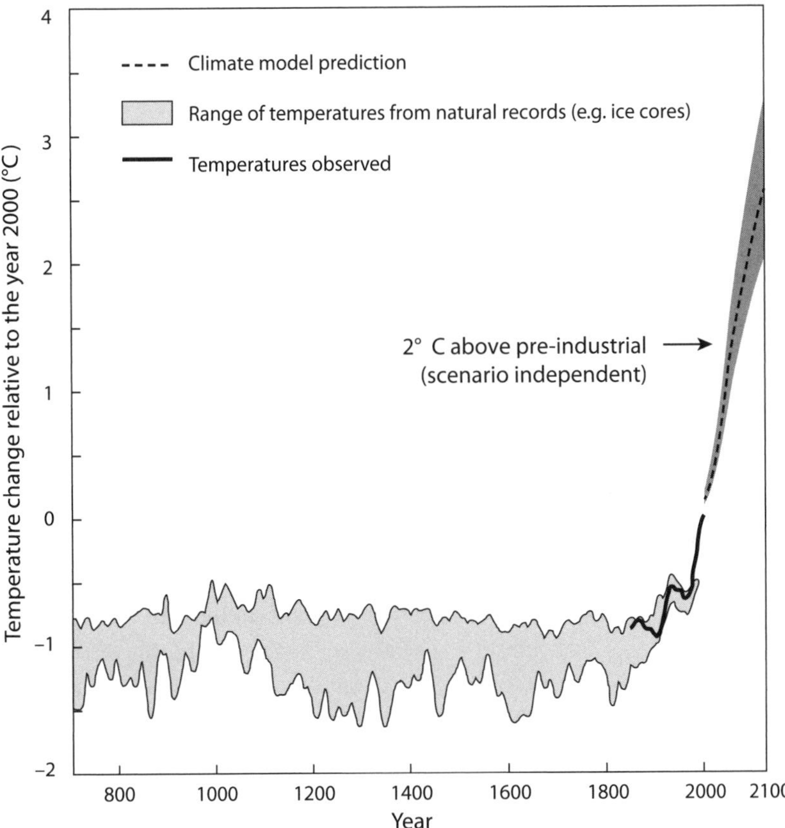

Figure 8.13. In the United Kingdom, the Hadley Centre for Climate Change has made predictions of the temperature rise averaged over the planet: the solid line shows a single simulation of warming over this century, and the gray shading indicates the uncertainty in that forecast. These predictions use highly simplistic feedback models that are the subject of much current debate and research. © Crown Copyright, Met Office.

run; this gives so-called physics ensembles, and allows estimation of the effects of varying the physics in each model.

At present, one decade into the twenty-first century, the key issues in climate modeling are being debated alongside much research. The science is focusing on the critical feedbacks in the Earth-air-ocean-ice-biosphere system. Certain climate simulations suggest that the varying higher amounts of moisture in the atmosphere in the middle decades of the twenty-first century—in particular, water in convective rainstorms—alter the extent of the ice coverage, the average sea levels, the flooding events, and the aridity of many countries. But these changes are in turn influenced by many other delicate changes, and sorting all this out presents a great intellectual and practical challenge.

Most importantly, what should we focus on doing now to help planners devise strategies to improve the quality of all life on planet Earth by the end of this century? We certainly need to understand the long-term weather better. Perhaps the major beneficiaries of the modern global economy could invest more in international centers for research focused on climate change—its science, and its mathematical computer modeling—to help answer some of these uncertainties.

The Fabric of Our Vision

In the previous sections we outlined some of the practical and theoretical issues that confront weather prediction and climate modeling on supercomputers. We outlined various developments, from increasing computer size, to improving the software, to the growing fleet of satellites gaining new data, which all help improve forecasting. And much more detail is necessary, especially about the moisture processes: details as to how the water evaporates from the oceans, rivers, lakes, and marshes, and from the trees, grasses, and so on, and how water condenses (and freezes) to ultimately form the multitudes of clouds that we watch drifting over us. Making further improvements to the software that represents physical processes such as these, together with better representations of the land and sea surface effects and the upper stratosphere, will all help in more accurately and reliably predicting events from local rainstorms to the gradual change of the climate.

There are always pressures to improve our ability to forecast, from tomorrow's weather to the changes in the climate, and to do it now! The decisions that governments take cost many billions of dollars and may have very serious consequences, for instance, on future flooding of low-lying cities and countries. A computer program that reliably tells us the consequences of society's actions would seem an urgent priority. Is reducing atmospheric carbon dioxide more effective than changing atmospheric methane, or something else, in ensuring that society achieves its goals—whatever they might be? Changes to the water cycle will probably turn out to be of more importance to life than changes to the average temperature, so it becomes important to know precisely how these are linked.

Because the only language that a computer speaks is mathematics, and because answers are required now, have we got as much as possible out of mathematics in the present use of supercomputers to model the evolution of weather? As we saw in figure 1.14, weather, or climate, is described by using weather pixels attached to a large grid, and all possible states of these pixels make up the weather state space. Each time we take a different computer, a different grid, and a different version of the rules to advance each weather pixel over time, a different model of the weather state space is obtained. Then, when the program is switched on, a different life history of computer weather forecasts is produced.

To predict the future, our aim each time is to follow the life history of planet Earth weather with a computer life history, or "movie," for the sequence of weather pixels. But in any such choice of weather state space there are more logically possible weather forecasts than there are particles in the universe. The first major challenge is to get the computer to follow these weather life histories in the planet Earth part of the weather state space, and not to drift into planet Venus weather, or other logical possibilities. The QG theory of chapters 5 and 6 tells us how to begin identifying the blue planet's weather by using hydrostatic and geostrophic balance. However, these forecasts always sensitively depend on various errors and inadequacies in our knowledge. This sensitive dependence usually destroys precise predictability in weeks, as we saw in chapter 7.

One of the ways that the rate at which errors undermine predictions can be slowed down is by implementing the conservation laws behind the scientific rules formulated in the tech boxes of chapter 2, but this remains to this day a challenging issue. As we described earlier, progress has been made with the latest generation of programs. In chapter 7 we described various transformations that focus the weather pixel evolution in the regions of weather state space that respect the conserved quantities: of mass (using density), energy (using potential temperature) and PV.

The exact life history of the weather that we live in is an idealization. An often-quoted saying, attributed to the statistician George Edward

Pelham Box, is "essentially all models are wrong, but some are useful." The best that we can do is compute a model life history that captures as many features of the weather as are of interest to us. This is directly analogous to an artist creating a painting—what we see on the canvas is not the same as the reality outside, as shown in figure 8.14. Our brains and imagination help to create the illusion of reality when viewing pictures and movies.

Figure 8.14. Weather pixels model the atmosphere just as an artist models the rural vista through the window. The point is that models are not the same as reality but should capture what interests us. René Magritte, *The Human Condition*, 1933. © 2012 C. Herscovici, London / Artists Rights Society (ARS), New York and Gianni Dagli Orti / The Art Archive at Art Resource, NY.

In the first two sections we described the modern techniques of ensemble forecasting and data assimilation that are being used to fill in various gaps in present-day knowledge of the weather. The geometry of constraints, scalings, and transformations from chapter 7 helps identify the more "realistic" computed life histories from those that belong to other worlds or even dramatic fiction. Will being more careful about the data and the physics, as we described earlier in this chapter, together with using bigger computers ever be sufficient to achieve our goal of accurate weather prediction? Our opinion is that it will always be necessary to get the most out of mathematics to improve our view of weather, and it will remain necessary to analyze mathematics further in order to get the latest improvements into the computer algorithms. Weather forecasters are continually seeking to improve the rules that direct the computer calculations. The ideal rule would predict all the weather behavior that is of interest while not taking very long to make that prediction. Scientists are continually experimenting using mathematical and computer simulations of simpler models, such as QG theory, to better formulate the pixel algorithms.

The next guiding hand that we mention from the underlying mathematics holds sway in the tropics because here Charney's state of geostrophic balance weakens to the point that it no longer controls the horizontal winds. Instead we have enhanced rainfall in the much warmer tropical zone, mainly in the huge tropical hot towers of convection usually associated with very large cloud structures. This daily movement of water from the ground up through the atmosphere releases large amounts of heat as the vapor condenses. Then it rains heavily. This water cycle moves the Sun's heat energy from the ground and sea surfaces to the upper atmosphere and eventually away from the equatorial regions.

In middle latitudes the westerlies ceaselessly carry the Bergen School's string of cyclones in an eastward direction around planet Earth. In the tropics the gentler surface trade winds occasionally give way to the fury of tropical cyclones or hurricanes. In most of the tropical zone, the atmosphere adjusts itself so that temperature surfaces are very nearly horizontal. This causes more significant but local vertical movement of air together with a local horizontal adjustment. The net effect of all these local tropical events is to move significant heat energy away from the equator, at higher altitudes in the troposphere. So the Gordian knot is cut a little differently in the tropics.

Figure 8.15. Swirling motion dominates the cloud patterns in midlatitudes, but the tropics are different. The tops of great towering convective storms can be seen along the equatorial region. © NEODAAS / University of Dundee.

The Bergen School had observed in the weather data of the middle latitudes that cyclones seemed to be born and to grow or develop on a temperature discontinuity known as the polar front (see figure 3.13), which separates warmer subtropical air from cooler polar air. Even though subsequent post-1950s theory suggests that cyclones might actually cause much of this frontal behavior, the presence of fronts raises one more challenge that we have not yet discussed: the issue of how to maintain the identity of a single quantity, such as a parcel of air, when that quantity evolves and partially extends across several pixels. In chapters 2 and 3 we saw Euler's generalization of Newton's laws of motion to describe the motion of continuous fluids such as air and water. Euler thought of the motion of parcels of fluid in which the parcels are more like smooth balloons than ragged clouds, and he realized that he could use the D/Dt operator to express the balloon motion of the fluid parcel in a fixed, ground-based, and grid-based way.

This change of viewpoint—from that of focusing on the identity of an object, where it is, and where it is going, to focusing on a list of what is at the background measuring stations—has a very simple analogy in the evolution of a modern notation in chess. The older notation in chess focused on the pieces and said where each piece would move to. This is efficient for a human brain because we know that all the other pieces stay in place as one piece moves. We used D/Dt to give this evolution in chapter 2, and we called this the parcel derivative because it follows an identified parcel of fluid. Modern chess notation instead focuses on a list of positions on the board. Because a chess board has only sixty-four squares or grid points on which there can be at most one piece, and because pieces do not magically appear, this is more efficient for computers although less so for humans. Human brains then look at the positions and interpret how the game is progressing.

Similarly, the modern notation for changes to pressure, moisture, temperature, and so on focuses on these weather pixel quantities at each position on a grid around the planet. Consider a parcel of air marked out by fine dust that originated in the Western Saharan Desert, as shown in figure 8.12, where the dust will take months to settle out. As the air parcel moves across a fixed set of pixels, how do we maintain its identity?

In the case of a chess game, we exchange or move a finite number of discrete objects around on the board as the game progresses. There is no

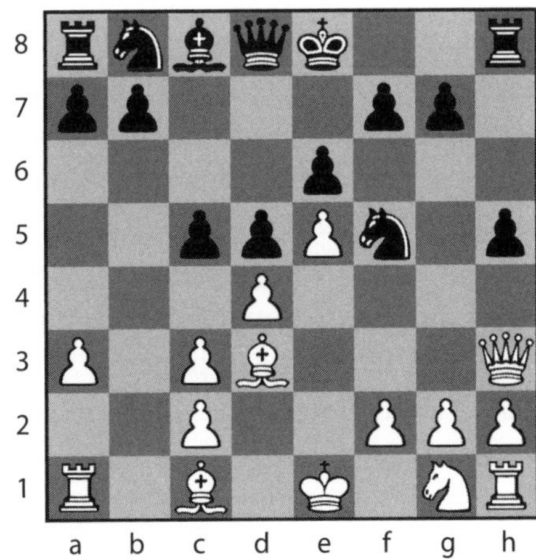

Figure 8.16. We can indicate the movement of the white knight by either identifying the piece and where it moves or by describing the entire board after the move. A computer would list the positions and what piece occupies each position after each move. There is a lot of repetition in this method; human brains then look at the differences to identify what happened. Computers model weather evolution using lists of positions and weather states, which can lose sight of the original identity of the air masses and what they are doing.

ambiguity about the identity of pieces. But we cannot do this with the weather pixels. We could actually think of the chess pieces as representing parcels of air, each with its own identity. But as time passes, these air masses move and deform and even mix with air from another pixel—in fact, usually each pixel can no longer be identified with any previous air mass. So again, how can air mass identity be respected? This is one of the key ideas behind the introduction of the SL and conservation methods that we described earlier in this chapter.

As figures 7.21 and 7.25 suggest, actual weather evolution can stretch and mix air masses. While this might not matter over a daily forecast, it can seriously harm predictions of moisture transport in the longer term. Careful solution of the scientific rules of chapter 2 applied to various idealizations of weather, much like Rossby's and Eady's methods described in chapters 5 and 6, shows that new types of solutions frequently appear that simulate the conceptual weather fronts of meteorology. These weather fronts typically occur when a warm air mass travels as a coherent package with not very much mixing but ends up next to (and mostly above) a much cooler air mass, usually with significantly different moisture levels. Keeping tabs on the two different air masses presents a real difficulty for modern computer models based on fixed grids. As storms develop, the stretching and mixing occur in a carefully orchestrated three-dimensional manner. So rapid changes may develop at many different length and time scales. Strong gusts are often experienced before it rains from a cold front; up and down drafts buffeted and destroyed the early balloon flights, as we saw in chapter 5.

The mathematics of parcel-based derivatives versus grid-based derivatives is straightforward when the quantities being differentiated are smoothly varying, but such mathematics can be tricky when not. Thus, there is another somewhat hidden constraint in the computer model of weather state space—how to respect identities of air parcels that do not naturally conform to the weather pixel setup on the computational grid and that may even partially mix when the quantities are not smoothly varying. The future generations of weather prediction programs will take more care of the dual structures of fixed pixels and moving air parcels. Conventional total, or parcel, derivatives will be calculated more accurately.

Most importantly, surfaces that separate distinct air masses—say, a hot and dry region from a cold and wet one—can move, develop, and disappear as the weather event unfolds. Air mass identity and evolution, even where there are sudden transitions of temperature and moisture, need to be maintained within the weather pixel rules. Delicate surfaces of discontinuity often occur at the edge of rain bands, as we see in figures 8.9 and 8.17, and this makes the already-difficult issue of forecasting precipitation even more challenging.

So we observe that computer models based on smoothly varying initial conditions, which are well represented on their pixels, often predict that the weather will evolve into phenomena across which warmth and moisture suddenly change in value (see the virga in figure 8.17). Such phenomena may not be well represented by the pixels. Representing discontinuities is not simply a technical problem for the computer models. We need to think about solutions in a radically new mathematical way. Again, the individual air parcel's journey pays no heed to the way the pixel description is chosen. Present-day time-stepping computer programs approximate the parcel's path within each time step but assume

Figure 8.17. *Cirrus uncinus* (hooked cirrus) and trailing virga photographed over Newbury, Berkshire, England, at midday on September 18, 2010. A computer model based on averaging the basic variables across its pixels would represent this type of phenomenon with only moderate success. Image © Stephen Burt.

that the heat and moisture are changing more smoothly in both space and time as the path is followed. How best to represent moving surfaces of discontinuity is a further challenge for contemporary research.

Although more than two millennia have passed since Aristotle first introduced "meteorology," weather is still something of an enigma; perhaps this explains why it remains a favorite topic of conversation! Weather often affects our frame of mind and our outlook: our reaction to its ceaselessly varying "moods" is intimate and personal. But these days we can also detach ourselves and take a much more global, top-down view of the atmosphere in motion. Images from satellites reveal the big-picture—from tropical thunderstorms and hurricanes interrupting the trade winds, to the stormy cyclones of middle and higher latitudes interrupting spells of fine weather. But any one image—spectacular as it might be— does not tell us where and when the next thunderstorm will develop or whether drizzle will turn to freezing rain and cause havoc on our roads.

To figure out where and when weather will change its mood, we need mathematics: we need to appeal to what lies invisible amid the beauty, the power, and the enigma of weather. The laws of physics must be encoded in a computer model. The latest observations and measurements need to be assimilated in just the right way, and the equations of motion, heat, and moisture need to be integrated so that the overarching conservation principles are respected. When this is all done, we would like to estimate how much faith can be placed on the forecast. Even further, we would like to evaluate Earth's future climate.

Significant improvement to forecasting relies on cooperation. Small armies of engineers and scientists continually improve the technology to observe and extend our knowledge of atmospheric processes. Then this growing body of knowledge needs to be converted into the language of mathematics and incorporated in the ever-more powerful computer models. Our message is that the smart use of mathematics gets the best out of these programs and helps us to see the ever-changing weather patterns emerging from the mists of chaos.

POSTLUDE

Beyond the Butterfly

Astronomy and meteorology have always played important and interactive roles in driving the historical development of physics and mathematics. Astronomy was among the first of the sciences to benefit from advances in mathematics: Newton's laws of motion and gravitation, supplemented by the conservation laws of energy and angular momentum, enabled the orbits of the planets to be evaluated for many years into the future. By the end of the eighteenth century, confidence in the science was captured by intricate models, such as the orrery, shown in figure Po.1.

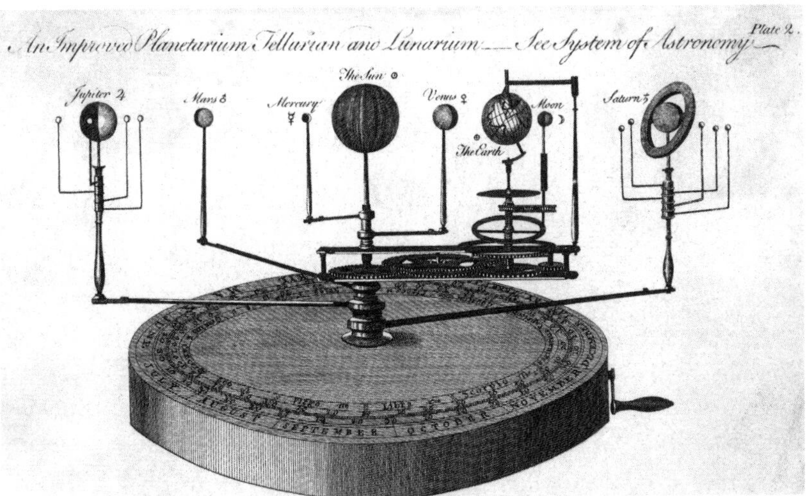

Figure Po.1. An orrery: a clockwork model of the solar system that epitomizes the Newtonian view of a deterministic universe. With knowledge of where to start, we can always say exactly what the future positions of the planets will be. © Bettmann/CORBIS.

Contrast this with the challenge of creating computer models of weather and climate where chaos plays a role, from the unpredictable swirls of smoke that blow away from chimneys, to the paths of tornadoes and hurricanes. How can we find those aspects of weather that are predictable? Our story has described how mathematics plays a vital role in answering this question.

This book has been about the role of mathematics in explaining *why* we are able to understand weather and climate, even in the presence of chaos. But our story is unfinished. Mankind is faced with the challenge of understanding and predicting how the component parts of the Earth system—the atmosphere, oceans, land, water, and life—interact and impact upon each other. Nonlinear feedback is ubiquitous in the Earth system illustrated in CI.15. But the good news is that the techniques developed over recent decades for analyzing and predicting weather, such as sophisticated computer algorithms for the transport of moisture and

Figure Po.2. This picture was taken from the International Space Station on July 31, 2011. Instruments carried on satellites allow scientists to observe the constituents of our atmosphere. Incorporating this information into computer simulations enable us to investigate how the different gases and aerosols will impact on climate. Our story has explained how mathematics that was originally developed for very different purposes, such as studying the ether or the dynamics of the solar system, is now helping us to understand the dynamics of the atmosphere and oceans, and the changes in our climate. Photo courtesy of NASA.

novel methods of data assimilation, are now being incorporated into simulations of the Earth system.

From time to time we may feel content to shrug our shoulders and claim that the flap of a butterfly's wing can undo all that a weather forecast seeks to rationalize. Yet weather—and, indeed, our Earth system—is not so easily disturbed. There are many detailed interactions, and there are degrees of unpredictability. But there are also many stabilizing mechanisms and, most importantly for understanding and prediction, there is the math to quantify the rules. For the past fifty years, math has played an ever more important role in designing better computer algorithms. The challenge for the future is to continue to develop, interpret, and use this science, technology, and mathematics in the most effective way.

Just as our quest to determine the behavior of the solar system produced surprises that led mathematicians to develop qualitative techniques for studying chaotic systems, so we might imagine that our efforts to understand the Earth system will lead to new mathematics and to a deeper appreciation of the world around us.

GLOSSARY

Air parcel: A small volume of air to which we assign values of the basic variables.

Basic laws: A set of physical laws written in mathematical form that describe how the basic variables interact with each other. (See tech box 2.3.)

Basic variables: The variables that describe the state of a small parcel of air; that is, temperature, pressure, density, humidity, wind speed, and direction. Knowing these defines the weather.

Circulation: The measure of the effective movement of air parcels around an imaginary loop in the fluid.

Computer program: A set of instructions that organizes a calculation of the values of the basic variables, given background data on the properties of the atmosphere and the basic laws.

Computer weather prediction: Use of a computer program written in terms of the basic variables that represents the basic laws.

Constants: Values of physical quantities that remain fixed while weather changes. Constants give the physical properties of the atmosphere; for example, the temperature at which water droplets freeze at a specified pressure.

Convergence: The apparent disappearance of air when its movement is viewed in a horizontal layer; the (usually) small amount of air that disappears actually moves vertically away from the layer.

Discontinuity: A sudden change in a variable either as we vary time or vary position; for example, a rain front, where the onset of heavy rain occurs in a few seconds.

Geostrophic balance: The horizontal wind blowing around the pressure isobars at a given height, which reflects the dominance of the Earth's rotation on the hydrostatically balanced air movement.

Gradient: The rate of change of a specified variable (such as temperature) in the direction that it maximally increases.

Hydrostatic balance: The layering of the Earth's atmosphere, which reflects the dominance of the effect of gravity (buoyancy forces) in the vertical.

Physical laws: Principles that always hold at any stage in the evolution of weather; for example, conservation of water mass: even though water changes its form as it moves around with an air parcel (say, changing from vapor to cloud to snowflakes), the total amount of water involved can be measured and accounted for.

Potential temperature: A transformed temperature (using the local pressure) that measures the heat energy in a parcel of air, no matter at what pressure the parcel is. (See tech box 5.2.)

Potential vorticity: A measure of the vorticity of a parcel of air normalized by a suitable gradient of the potential temperature. (See chapter 5 and tech box 7.1.)

Variables: Values of physical quantities that may change as weather changes; for example, the rainfall rate over central Seattle.

Vortex: An idealized eddy, or whirlpool, in a fluid; vorticity and circulation are ways to measure its strength of rotation.

Vorticity: The measure of the rotation of air in a particular direction.

BIBLIOGRAPHY AND FURTHER READING

Books: General

The following titles are accessible to a general readership and cover aspects of the development of meteorology that we have been concerned with:

Ashford, O. M. *Prophet or Professor? Life and Work of Lewis Fry Richardson*. Adam Hilgar, 1985.

Cox, J. D. *Storm Watchers: The Turbulent History of Weather Prediction from Franklin's Kite to El Niño*. Wiley, 2002.

Friedman, M. R. *Appropriating the Weather: Vilhelm Bjerknes and the Construction of a Modern Meteorology*. Cornell University Press, 1989.

Harper, K. C. *Weather by the Numbers: The Genesis of Modern Meteorology*. MIT Press, 2012.

Nebeker, F. *Calculating the Weather: Meteorology in the 20th Century*. Academic Press, 1995.

The next four publications cover many of the developments in theoretical meteorology and weather forecasting we have described, but at a more technical level:

Lindzen, R. S., E. N. Lorenz, and G. W. Platzman, eds. *The Atmosphere—A Challenge: A Memorial to Jule Charney*. American Meteorological Society, 1990. [Contains reprints of Charney's papers.]

Lorenz, E. N. *The Essence of Chaos*. University of Washington Press, 1993.

Lynch, P. *The Emergence of Numerical Weather Prediction: Richardson's Dream*. Cambridge University Press, 2006.

Shapiro, M., and S. Grønås, eds. *The Life Cycles of Extratropical Cyclones*. American Meteorological Society, 1999.

The following three books are an excellent introduction to some of the basic physics relevant to our story, and they are accessible to a wide readership:

Atkins, P. W. *The 2nd Law: Energy, Chaos and Form*. Scientific American Books Inc., 1984.

Barrow-Green, J. *Poincaré and the Three-Body Problem*. American Mathematical Society, 1997.

Feynman, R. P. *The Character of Physical Law*. Penguin, 1965.

Books: Technical

The following texts are for graduate-level students:

Gill, A. E. *Atmosphere-Ocean Dynamics*. International Geophysics, 1982.

Holm, D. D. *Geometric Mechanics. Part 1: Dynamics and Symmetry.* Imperial College Press, 2008.

Kalnay, E. *Atmospheric Modelling, Data Assimilation and Predictability.* Cambridge University Press, 2003.

Majda, A. J. *Introduction to PDEs and Waves for the Atmosphere and Ocean.* American Mathematical Society, 2002.

Norbury, J., and I. Roulstone, eds. *Large-scale Atmosphere-Ocean Dynamics.* Volume One: *Analytical Methods and Numerical Models.* Volume Two: *Geometric Methods and Models.* Cambridge University Press, 2002.

Palmer, T. N., and R. Hagedorn. *Predictability of Weather and Climate.* Cambridge University Press, 2007.

Richardson, L. F. *Weather Prediction by Numerical Process.* Cambridge University Press, 1922 (reprinted 2007).

Vallis, G. K. *Atmospheric and Oceanic Fluid Dynamics.* Cambridge University Press, 2006.

Articles

The following review papers are aimed at a general readership:

Jewell, R. "The Bergen School of Meteorology: The Cradle of Modern Weather Forecasting." *Bulletin of the American Meteorological Society* 62 (1981): 824–30.

Phillips, N. A. "Jule Charney's Influence on Meteorology." *Bulletin of the American Meteorological Society* 63 (1982): 492–98.

Platzman, G. "The ENIAC Computations of 1950: The Gateway to Numerical Weather Prediction." *Bulletin of the American Meteorological Society* 60 (1979): 302–12.

Thorpe, A. J., H. Volkert, and M. J. Ziemiański. "The Bjerknes' Circulation Theorem: A Historical Perspective." *Bulletin of the American Meteorological Society* 84 (2003): 471–80.

Willis, E. P., and W. H. Hooke. "Cleveland Abbe and American Meteorology, 1871–1901." *Bulletin of the American Meteorological Society* 87 (2006): 315–26.

The following review paper is aimed at graduate-level students:

White, A. A. "A View of the Equations of Meteorological Dynamics and Various Approximations," in vol. 1 of *Large-Scale Atmosphere-Ocean Dynamics*, edited by J. Norbury and I. Roulstone. Cambridge University Press, 2002.

Classic Papers

Abbe, C. "The Physical Basis of Long-Range Weather Forecasts." *Monthly Weather Review* 29 (1901): 551–61.

Bjerknes, J. "On the Structure of Moving Cyclones." *Monthly Weather Review* 49 (1919): 95–99.

Bjerknes, V. "Das Problem der Wettervorhersage, betrachtet vom Standpunkte der Mechanik und der Physik" ("The Problem of Weather Forecasting as a Problem in Mechanics and Physics"). *Meteor. Z.* 21 (1904): 1–7. English translation by Y. Mintz,

1954, reproduced in *The Life Cycles of Extratropical Cyclones*. American Meteorological Society, 1999.

———. "Meteorology as an Exact Science." *Monthly Weather Review* 42 (1914): 11–14.

Charney, J. G. "The Dynamics of Long Waves in a Baroclinic Westerly Current." *Journal of Meteorology* 4 (1947): 135–62.

———. "On the Scale of Atmospheric Motions." Geofysiske publikasjoner, 17 (1948): 1–17.

Charney, J. G., R. Fjørtoft, and J. von Neumann. "Numerical Integration of the Barotropic Vorticity Equation." *Tellus*, 2 (1950): 237–54.

Eady, E. T. "Long Waves and Cyclone Waves." *Tellus* 1 (1949): 33–52.

———. "The Quantitative Theory of Cyclone Development." *Compendium of Meteorology, edited by* T. Malone. American Meteorological Society, 1952.

Ertel, H. "Ein neuer hydrodynamischer Wirbelsatz," Meteorologische Zeitschrift 59 (1942): 271–81.

Ferrel, W. "An Essay on the Winds and Currents of the Oceans." *Nashville Journal of Medicine and Surgery* 11 (1856): 287–301, 375–89.

———. "The Motions of Fluids and Solids Relative to the Earth's Surface." *Mathematics Monthly* 1 (1859): 140–48, 210–16, 300–307, 366–73, 397–406.

Hadley, G. "Concerning the Cause of the General Trade Winds." *Philosophical Transactions of the Royal Society of London* 29 (1735): 58–62.

Halley, E. "An Historical Account of the Trade Winds, and Monsoons, Observable in the Seas between and Near the Tropics, with an Attempt to Assign the Physical Cause of the Said Winds." *Philosophical Transactions of the Royal Society of London* 16 (1686): 153–68.

Hoskins, B. J., M. E. McIntyre, and A. W. Robertson. "On the Use and Significance of Isentropic Potential Vorticity Maps." *Quarterly Journal of the Royal Meteorological Society* 111 (1985): 877–946.

Lorenz, E. N. "Deterministic Nonperiodic Flow." *Journal of Atmospheric Sciences* 20 (1963): 130–41.

Rossby, C.-G. "Dynamics of Steady Ocean Currents in the Light of Experimental Fluid Mechanics." *Papers in Physical Oceanography and Meteorology* 5 (1936): 2–43 [Potential vorticity appears for the first time in this paper.]

———. "Planetary Flow Patterns in the Atmosphere." *Quarterly Journal of the Royal Meteorological Society* 66, Supplement (1940): 68–87. [Review bringing together the ideas set out in the 1936 and 1939 papers.]

———. "Relation between the Variations in Intensity of the Zonal Circulation of the Atmosphere and the Displacement of Semi-Permanent Centers of Action." *Journal of Marine Research* 2 (1939): 38–55. [Key paper on Rossby waves.]

INDEX